The Snake with Golden Braids

The Snake with Golden Braids

Society, Nature, and Technology in Andean Irrigation

Stephen G. Bunker

LEXINGTON BOOKS

A Division of
ROWMAN & LITTLEFIELD PUBLISHERS, INC.
Lanham • Boulder • New York • Toronto • Oxford

LEXINGTON BOOKS

A division of Rowman & Littlefield Publishers, Inc.
A wholly owned subsidiary of The Rowman & Littlefield Publishing Group, Inc.
4501 Forbes Boulevard, Suite 200
Lanham, MD 20706

PO Box 317
Oxford
OX2 9RU, UK

British Library Cataloguing in Publication Information Available
Library of Congress Cataloging-in-Publication Data

Bunker, Stephen G., 1944–
The snake with golden braids : society, nature, and technology in Andean irrigation / Stephen G. Bunker.
p. cm.
Includes bibliographical references and index.
ISBN: 978-0-7391-1197-0
ISBN: 0-7391-1197-3

1. Indians of South America—Agriculture—Peru—Huanoquite. 2. Indian cosmology—Peru—Huanoquite. 3. Indigenous peoples—Ecology—Peru—Huanoquite. 4. Irrigation engineering—Peru—Huanoquite—History. 5. Irrigation farming—Peru—Huanoquite—History. 6. Landscape assessment—Peru—Huanoquite. 7. Huanoquite (Peru)—Social conditions. 8. Huanoquite (Peru)—Environmental conditions. I. Title.
F3429.1.H8B86 2005
333.91'3'098537—dc22 2005016731

Printed in the United States of America

™ The paper used in this publication meets the minimum requirements of American National Standard for Information Sciences—Permanence of Paper for Printed Library Materials, ANSI/NISO Z39.48–1992.

Dedication

I thank first the earth, the water, the wind, and the landscape they formed for teaching me not just what questions to ask, but how to ask them.

Next, I am deeply grateful to the Huanoquiteños who, when they learned what I was trying to discover, taught me far deeper ways to see into a landscape.

Finally, I am especially grateful to the men who introduced me to their ancestors, and to their ancestors' gods, especially to the snake with golden braids.

Contents

Preface

The Snake With Golden Braids recounts my quest to understand how local inhabitants came to know the extremely complex and rugged landscape around Huanoquite, a small village about 40 miles from Cuzco, Peru, and then to build and maintain a system of irrigation ditches across it. Huanoquite is blessed with an extraordinarily favorable topographic anomaly. High on steep slopes that rise from the Ccorca river, and in stark contrast with the huge peaks and ridges that rise sharply above it, lies a gently rolling plain, or pampa, relatively free of rocks and quite fertile. The Huanoquiteños' ancestors located their village just above this pampa, and just below some smaller fields that, though not as flat as and less fertile than the pampa, slope gently enough to reward cultivation.

Nature's topographic gift to Huanoquite, though, was deeply flawed. Neither the pampa nor the fields had enough water to support agriculture. Several strong rivers and streams emerge high in the surrounding mountains—some by slowly accumulating small surface flows, and one as a fully formed and very powerful subterranean flow bursting forth from under a huge rock formation—but each then flows down its own deep, narrow canyon, bounded by long, steep ridges that keep its accelerating plunge to the Ccorca far to the west of the pampa and the small fields above it.

The Huanoquiteños' forefathers figured out how to solve this problem, and built the extraordinarily complex system of irrigation ditches necessary to do it, long before the Spanish conquest, and probably before the Inca one as well. The combination of flat land and abundant water that they created in this harsh and parsimonious terrain raised the region's carrying capacity well above that of the surrounding areas. The relatively dense human population that grew up once the ditches were working their magic, and the rich and varied production of crops and animals that their felicitous combination of land and water made possible, supported the construction of an extensive network of trade routes up from the Apurimac valley and on through to Cuzco and beyond. Tributary roads descended to these trade routes from smaller, less productive valleys and the little agricultural hamlets they supported. Other minor roads climbed to mountain grasslands too high, dry and cold to support agriculture but adequate to pasture llamas and other camelids. Highland herdsmen could exchange their flesh, wool, and utility as beasts of burden for crops cultivated below. Both sources of ancillary trade added variety and volume to the flow of goods and power through Huanoquite.

The obstacles that the designers and builders of these ditches had to overcome to create such a richly productive and powerfully commercial system from such starkly barren terrain were as imposing as the mountains themselves. The solutions they imagined and then implemented were conceptually, topographically, and architecturally so complex that even with my own considerable experience in exploring and surveying high mountains, and despite the obvious willingness of many Huanoquiteños to tell me what they knew, it took me months of walking the hills and interviewing their inhabitants to figure out where the first ditches had flowed and how they had been built. Indeed, I only realized that the original pre-Incaic system had existed at all after spending considerable time and effort observing, talking about, and measuring the trajectory, flow, and general condition of the far simpler, but still extremely complex, system that had been built many centuries later, under the direction of local *hacendados* who had inherited the huge tracts of land and rights to indigenous labor that their imperialist Spanish ancestors wrested from the communities they conquered.

The simpler, more recently built, system is still very much in use. The steep slopes and complex topography it traverses are sufficiently challenging, and the job of maintaining the ditches is sufficiently onerous, that I was first drawn to the question of how a community maintained a technology built by earlier generations into its own landscape. The question of how a community might lose an essential ancestral technology became more salient as I learned how much contention there was about the division of water and about the division of work, and as I heard about the resentment against the authorities who organized the ditch cleaning, and their complaints about the impossibility of mobilizing enough labor to maintain the ditches in good condition.

I had to observe closely the ways ditch maintenance was organized and carried out. Understanding how the ditches were used required me to become intimately familiar not only with the ditches but also with the landscape they traversed. It was this process that revealed the original ditches. While tracking the more modern ditches, I discovered some severely eroded slopes. People I talked to affirmed that construction must have been very difficult, and that repair and maintenance were particularly challenging, and critically important, in these sections, but they could not explain how the erosion had occurred.

When I climbed these very steep slopes to learn the answer myself, I discovered traces of what could only have been another ditch, much higher up. This ditch must at some point have burst its banks and channeled all of its water to the single, highly vulnerable, point where the erosion started. The intensely concentrated water flowing through such a breach carved into and destabilized the slope below. The later construction of the lower ditch across these unstable sections required considerable engineering skill and constant precautions against breaks and collapses. All of the ditch authorities I interviewed clearly felt that the repairs they were able to carry out with the labor they could mobilize were inadequate to assure the continued security of the water flow across these precarious sections, and my own observations confirmed that the walls were being built to minimal height and thickness rather than to an optimal strength.

I was intrigued to know why and how the older upper ditch had been so neglected as to lead to such devastating erosion. Its higher trajectory would have irrigated larger areas than the ditch that replaced it, so its abandonment had obviously reduced the agricultural wealth of the village. The people I asked about the old ditches responded that their ancestors had had powers that they did not, and so were able to repair and maintain ditches better than they could. When I pushed harder on the question of why there was no attempt to resuscitate the old ditches, I was told stories about a man who had killed a snake with golden braids. The murdered snake withdrew the water that filled the ditches back into the mountain, and leveled a curse against all Huanoquiteños—that anyone who attempted to channel water back into the ditches would be paralyzed and die.

I did not see then how these stories about super-human ancestors and curse-leveling reptiles could answer my questions, so I resolved to trace the trajectory of the old ditches back to their original source. To do this, I used modern surveying tools to follow the minimal possible downward slope, that is, the slope steep enough to flow but flat enough to maximize the area that could be irrigated. I learned to detect changes in vegetation that might indicate an earlier concentration of moisture. As I advanced in my explorations, I got better at recognizing faint traces of the remains of old retaining walls and channels.

As my explorations became increasingly fruitful, my original concern to understand how contemporary generations might retain or lose technologies and recreate landscapes that they inherited from their ancestors was subordinated to the awe-struck question of how members of what would have been a small, isolated community could have imagined, and then built, a system capable of carrying so much water so far across such a difficult landscape.

I had by then carried out very precise measurement and mapping of the trajectories, construction and maintenance of the ditches still in use. I had also conducted extensive interviews with members of the community. This work taught me a great deal about the landscape, about the soils and rocks it was made of, about the kinds of effort and organization that maintaining ditches in this landscape required, and about the conditions and the consequences of breaks in a ditch that was not adequately maintained.

I noticed that different members manifested very different degrees of knowledge and attitudes about the ditches they still used. Huanoquiteños tended to agree with each other, and with me, in their estimation of different individuals' knowledge and capacity. This meant that the individuals with the deepest understanding of and sensitivity to the environment, and knowledge of the technical and material processes needed to maintain the ditches in that environment, were the most able to mobilize their fellows to the work required.

As I turned some of these understandings to the far more complex job of learning how the first ditches had been imagined and constructed, I realized that my informants used very different language and very different mediums to talk about the old and the new ditches. The new ditches were described in terms of fairly conventional Western narrative, while the old ditches were described in terms of mystery and myth. I was at first too busy with the physical chore of learning and mapping where the old ditches had gone to fully appreciate these differences, but I did eventually notice that I was having a much easier time discussing the ditches, the landscape, and the myths about them when I was sharing the physical experience of moving along the ditches' courses, or exploring the parts of the landscape around the ditch relevant to its course.

Years later, as I returned to reconsider my first analyses of ditch construction and maintenance, I finally came to understand that this greater ease came from the fact that much of what I was learning was through my bodily experience of ditch and landscape, and that this bodily knowledge could be communicated and shared quite effectively, even across the boundaries of my dependence on an over-academic Spanish whose structure and assumptions were largely alien to the Quechua in which my informants knew and thought about landscape, ditches, and the spirits that regulated them. It was only after rethinking the Huanoquiteños' myths and legends, after re-imaging and re-feeling the rocks, water, animals, and spirits that the myths and legends had taught me about, and after realizing that my bodily experience of the ditches and of the landscape they crossed had to be combined with a physical as well as a metaphysical understanding of what these myths and legends mean, that I could make sense of what I had been told and of what I had felt and seen about the ditches during my field work.

During this same time, I began to see, and feel, the snake with golden braids as the animal manifestation of the ditches, of the water in them, and then increasingly of the understandings of landscape and water flow that had been bestowed on the builders of the original ditches. Thinking and feeling the snake showed me how the community that emerged around the ditches had granted consensual authority (cf. Geertz, 1959) over their maintenance and regulation to those individuals who could communicate with the snake.

These understandings drove me to read phenomenologists Husserl and Merleau-Ponty, who both wrote extensively about how we humans learn through our bodies, and about how the Cartesian models of thought that dominate Western science distort our understandings of knowledge by attributing all knowledge to the mind. I returned as well to Reichel-Dolmatoff's (1971, 1976) writings on Amazonian cosmologies about human participation in and obligations to the ecology of the places they inhabit and subsist upon. I saw much more clearly than I had in earlier readings how central to Reichel-Dolmatoff's vision were communications with and learnings from animals, plants, and rivers. I then discovered the fascinating literature on Amazonian ways of knowing and learning from animals (Descola, 1994; Arhem, 1996; Rival, 1996; Hornborg, 1996, 2001) together with more general discussions in Feld and Basso (1996) of the ways that movements, sounds, and other sensual experiences create social knowledge and meaning of place, and shape the indigenous ways of knowing that comprise local cosmologies.

Most striking in this literature, and most relevant to what I had experienced in Huanoquite, was Tim Ingold's (1996: 40) explanation of the integration of observation, imitation, and bodily practice that a novice hunter must learn from his environment and the animals in it; "…to observe is actively to attend to the movements of others; to imitate is to align that attention to the movement of one's own practical orientation towards the environment." Ingold neatly captures the intersection between 1) the phenomenological critique of Cartesian logic and the primacy it attributes to mind in all cognitive processes and 2) the ethnographic recognition that there are other, non-Cartesian ways of knowing, which understand humans as only one of multiple sentient, intentioned life forms that constitute the natural world. Humans are therefore subordinate to, rather than dominant over, the abstract laws and the material processes of the landscape—and to the hydraulics, the climate and the species-diverse ecosystemic dynamic that constitute the environment they live in and depend on.

Combining the Huanoquiteños' stories of the snake and of other animals with the knowledge I gained from my kinetic, respiratory, muscular, visual, and gravitational experience of the landscape and of the flows of water across, around, and down its rocks, soils, and vegetation enabled me to imagine how particularly sensitive Huanoquiteños might have learned much more than I, with my far less sensuous experience of the locality and the impediments of my Cartesian training, could from animals', earth's, and water's movements. I could begin to imagine as well how these ancients imitated and incorporated those movements into their own bodily practice, and how awareness of their own body practice formed their own personal life-narratives—both in relation to society and and as a manifestation of the regularity of material process in their environment. They could, then, collectively and over time, abstract their knowledge of bodily practice into a common frame of understanding and rules of right behavior for the ditch-dependent community as a whole.

The snake, and other animals, thus guided the ditch-builders' learning of special bodily/sensual knowledge. Popular recognition and appreciation of this special knowledge and of the community's dependence upon it validated the consensual, or shamanic, authority (cf. Geertz, 1959; see also Reichel-Dolmatoff, 1976). The conquering Spanish found this kind of authority so alien, and so threatening, that they would call their murder of those who held it an "extirpation of idolatries."

Thus, as I re-imagined and rethought the combination of kinetic experience, legends about the snake with golden braids, observation of the ditches and the movement of water along it, and as I used these experiences to feel and follow the snake's movements across the landscape. I saw that the builders of the original ditches could understand their landscape, and the ecological relations within it, by observing and imitating the snake, and the ways the snake's movements replicated the waters. That community granted them special authority because it recognized the importance of their knowledge for organizing the maintenance of the ditches on which life and prosperity depended and for maintaining right relations with the environment.

This book follows that intellectual and spiritual journey, and recounts what it taught me, not just about building and managing the ditches, but about the very different ways of knowing the environment and the spirits of the earth that had informed and empowered their original builders. I relate the multiple means that I combined—from absolutely positivist surveying measurements and carefully controlled interviews to hermeneutical interpretations of my own bodily understanding of legends and landscapes—to unravel the mystery of how the ditches were imagined, built, and maintained, but I also examine what the ways I learned about the ditches tell us about how a society's social and natural histories, its cosmologies, its relations to its own environment, and its interactions with other life forms are written in the landscape from which it draws subsistence and its own sense of beauty.

The first four of the photos that follow show one of the most illuminating moments in my physical search for the snake with golden braids. My compadre Demetrio Pantoja, who was among the men most responsive to my questions about the ditches, invited me one day to accompany his nephew José on a trip into the mountains to conduct some business with a herdsman there. We rode horseback, higher and higher into the Abra de Vilcacunca, and then along high ridges that looked down into the Apurimac Valley's descent to the Amazon. After meeting with the herdsman, José led me to the base of a hill so steep that we could only continue on foot. After a mighty climb, we crested out onto a wide plateau that held several large lakes.

The spirit of the place filled me with such excitement and delight that I couldn't stop walking and looking, seeing the lakes from different perspectives. I was drawn to a huge rock that rose above them. It was only later, when I was calmer, that I realized that the great rock (in the third and fourth photos) dramatically resembled a bearded human visage—which in Andean cosmology would be Viracocha, the god who most directly created and controls mountains, water, time, and space. José then told me that the lakes, and the rock, all named Qaranqa, were the source, and the protector, of the water that emerges, over 20 kilometers away and almost 600 meters lower, to form the Qewar River, the primary source of Huanoquite's irrigation water. He said that a priest had thrown *cantuta* flowers in one of these lakes, and that they had later emerged below. I later discovered that Qaranqa was as well the home, and the protector, of the snake.

José took the second photo shown, which shows me standing above one of the lakes. I took the rest. The last picture shows a group of Huanoquite's men at a *faena*, a community ditch cleaning and repair in which all users are supposed to assist. I include this picture to show the extremely rugged terrain in the background, the source of irrigation water and the obstacle to its management. The maps that follow the photographs show both the contours of the drainage basin and the trajectories of the ditches. The transect below the first map portrays the relief from point A to point B marked on the map itself. The same letters identifying each ditch on the maps will appear again in the textual description of each ditch in Chapters Three and Four.

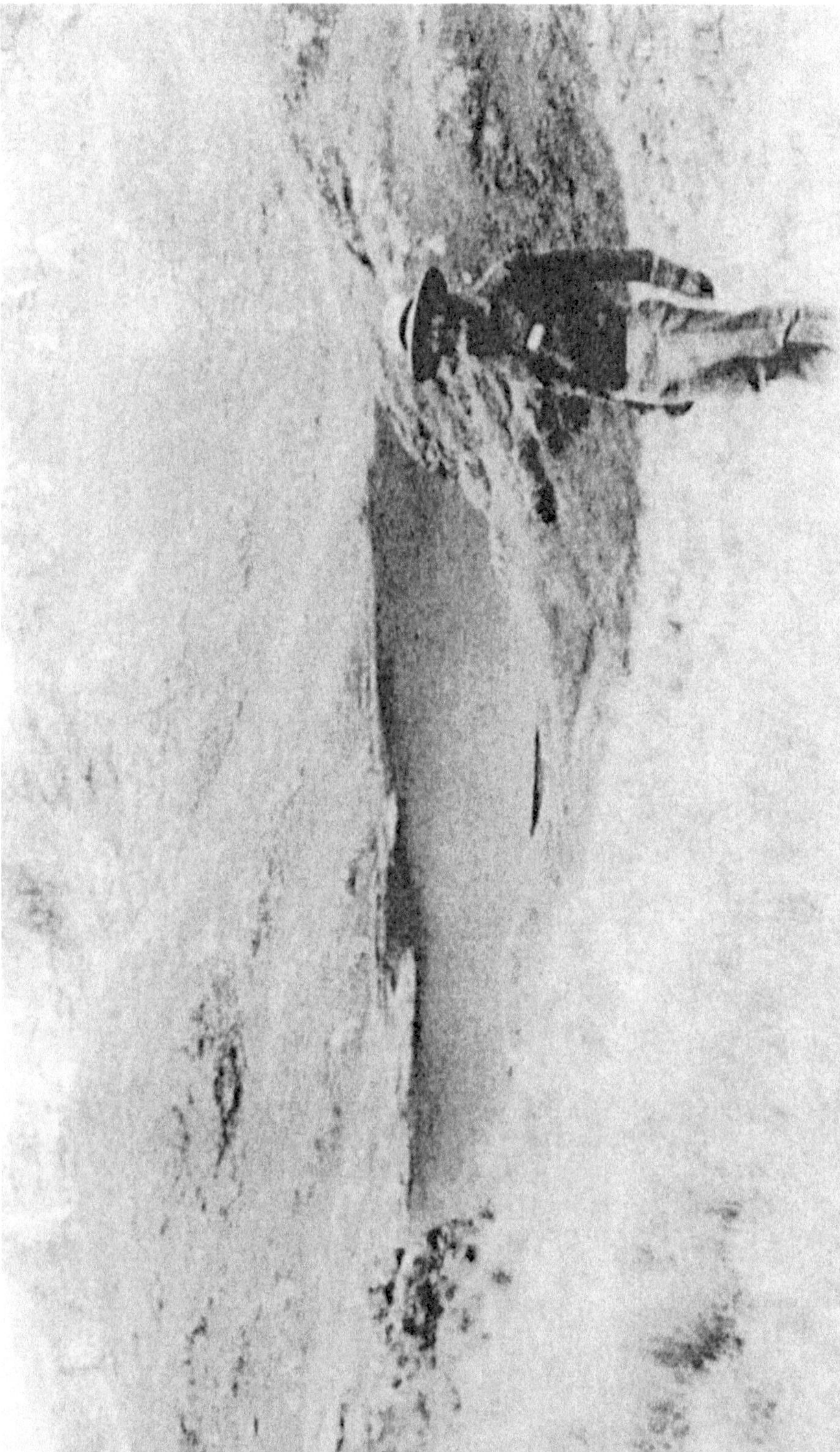

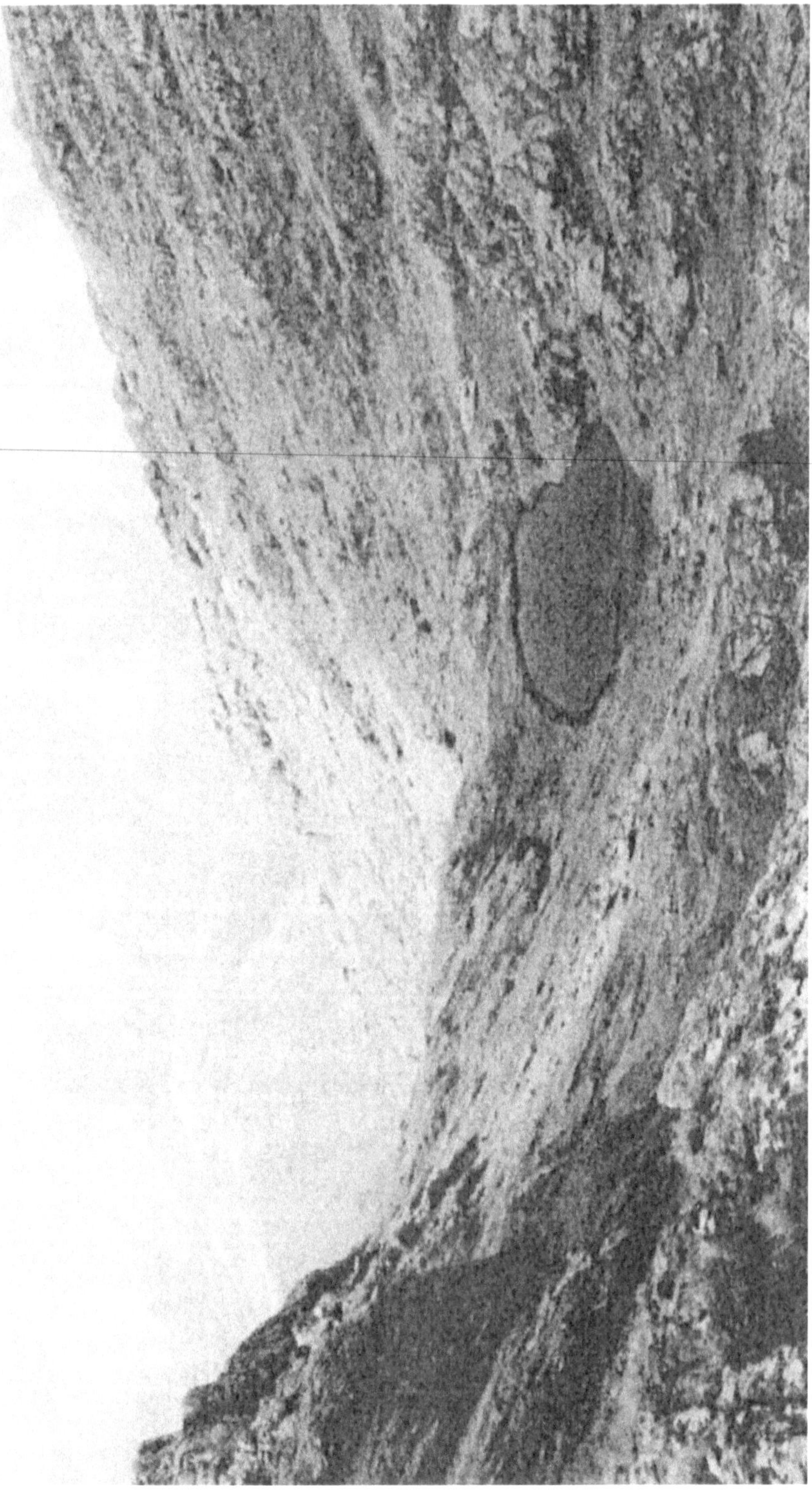

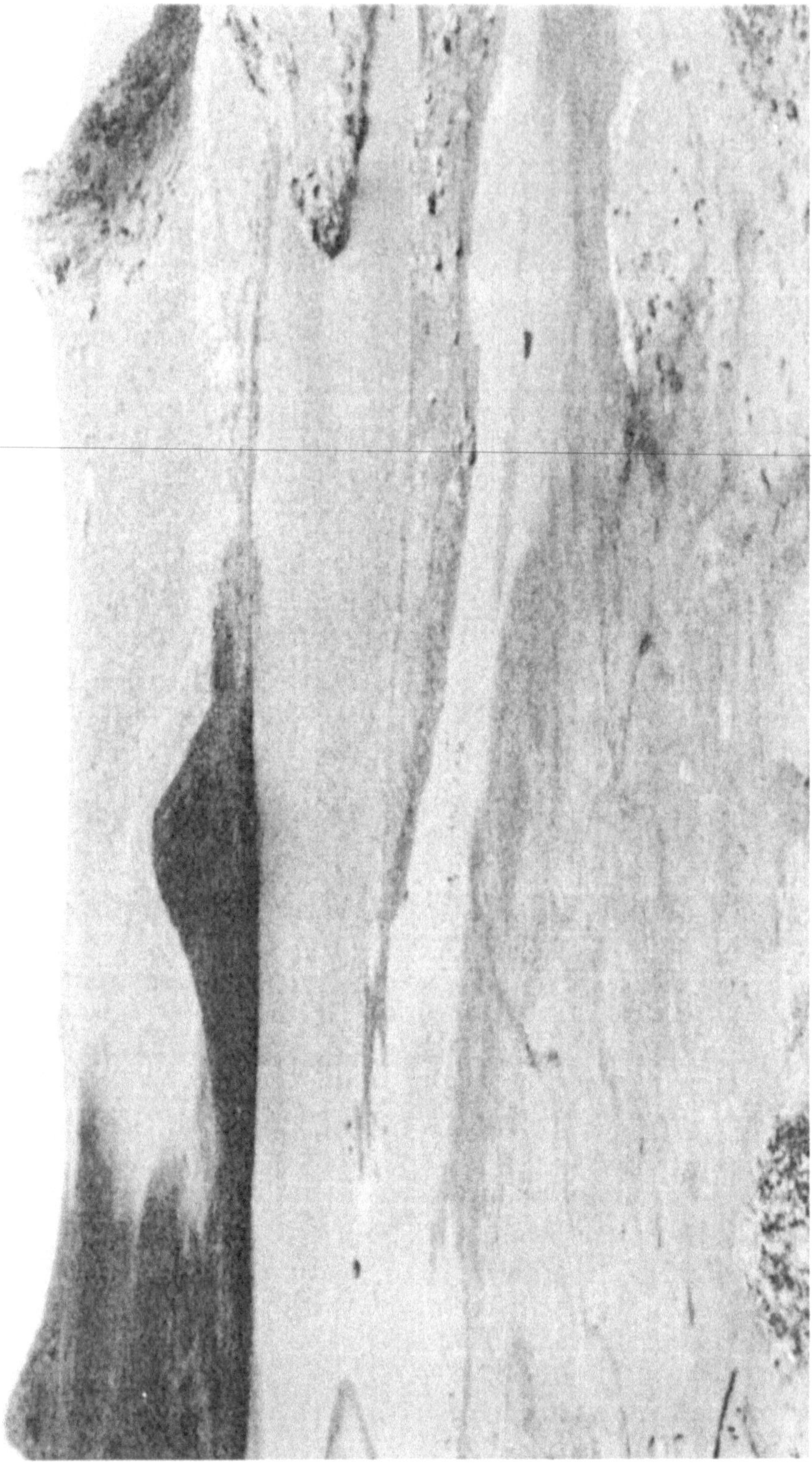

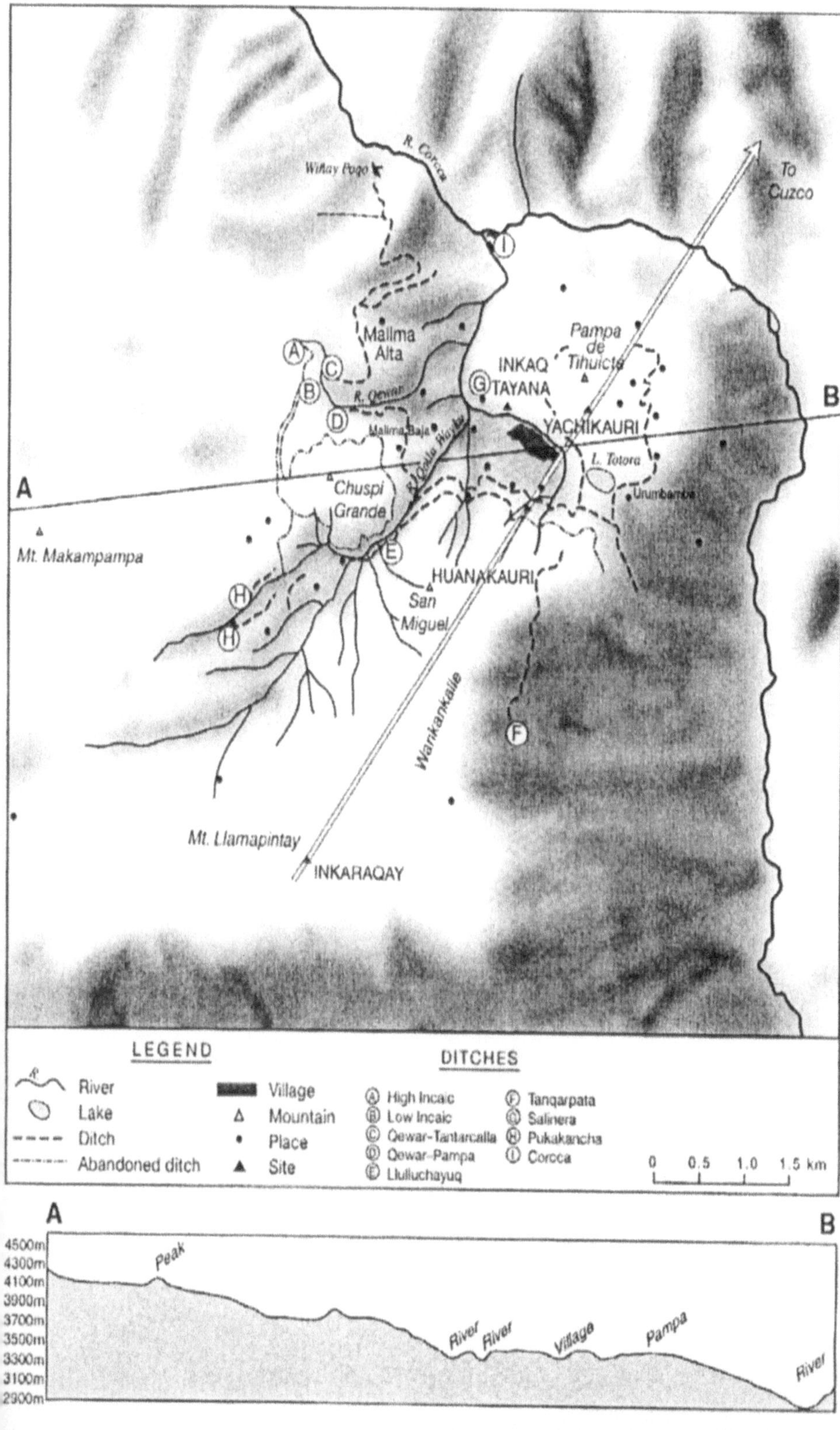

R. Corcca
Wiñay Poqo
To Cuzco
Mallma Alta
R. Qewar
INKAQ
TAYANA
Pampa de Tihuicte
B
YACHIKAURI
Mallma Baja
L. Totora
Urumbamba
A
Chuspi Grande
Mt. Makampampa
HUANAKAURI
San Miguel
Wankankalle
Mt. Llamapintay
INKARAQAY
LEGEND
River
Lake
Ditch
Abandoned ditch
Village
Mountain
Place
Site
DITCHES
A High Incaic
B Low Incaic
C Qewar-Tantarcalla
D Qewar-Pampa
E Llulluchayuq
F Tanqarpata
G Salinera
H Pukakancha
I Corcca
0 0.5 1.0 1.5 km
A
B
4500m
4300m
4100m
3900m
3700m
3500m
3300m
3100m
2900m
Peak
River
River
Village
Pampa
River

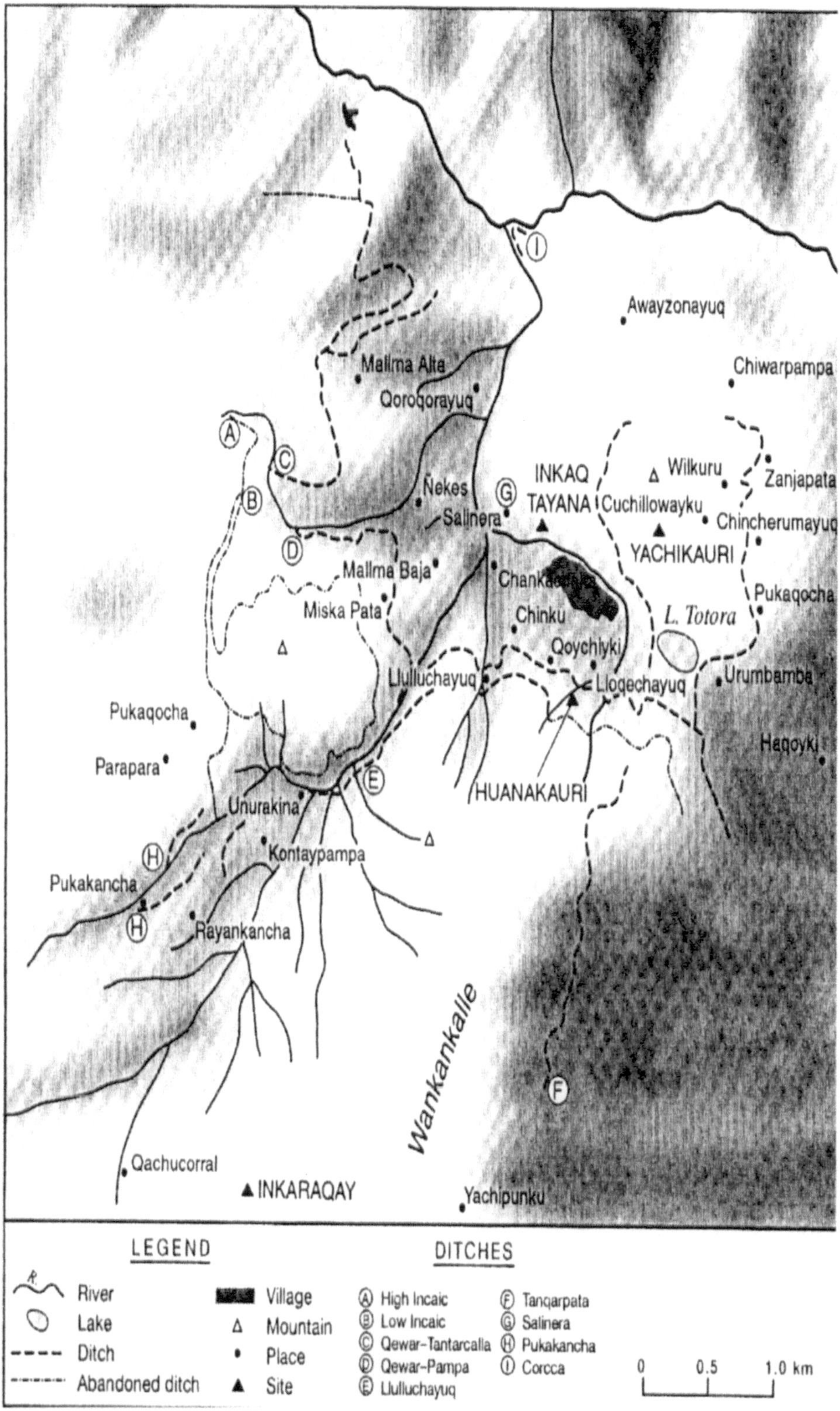

Awayzonayuq
Chiwarpampa
Mallma Alta
Qoroqorayuq
INKAQ TAYANA
Wilkuru
Zanjapata
Ñekes
Salinera
Cuchillowayku
Chincherumayuq
YACHIKAURI
Mallma Baja
Miska Pata
Chinku
Pukaqocha
L. Totora
Qoychiyki
Llulluchayuq
Llogechayuq
Urumbamba
Pukaqocha
Parapara
Haqoyki
HUANAKAURI
Unurakina
Kontaypampa
Pukakancha
Rayankancha
Wankankalle
Qachucorral
INKARAQAY
Yachipunku
LEGEND
River
Lake
Ditch
Abandoned ditch
Village
Mountain
Place
Site
DITCHES
A High Incaic
B Low Incaic
C Qewar-Tantarcalla
D Qewar-Pampa
E Llulluchayuq
F Tanqarpata
G Salinera
H Pukakancha
I Corcca
0
0.5
1.0 km

Acknowledgments

Jane Collins, Neil Whitehead, and Mara Loveman have been extraordinarily generous with their time and with their wisdom in commenting on this text as it evolved. Myron Eshowsky has inspired me importantly, and in ways that went beyond his already benevolent and loving intentions. My mother, Priscilla Fleitmann Bunker, and my brother, Paul Bunker, have commented on the work, and encouraged me in it, in deeply important ways. Linda Seligmann used her vast understanding of Andean societies to help me orient myself to the booming, buzzing confusion of field work, and has since most generously given me back some of my photographs from that time.

Chapter One

The Huanoquite Water Works: Environmental Sensitivity and Technological Innovation

Human activities often change the landscape in ways that subsequent generations take for granted as they react to, and use, what for them is the received environment. Deliberate modifications of the landscape may have required technologies, organization, and special skills and knowledge that are lost because subsequent generations no longer confront the environmental problems their ancestors solved, or because an exogenous system of social power disrupted and displaced the community their ancestors formed. Unintentional modifications, on the other hand, may damage the environment in ways that restrict human production capacities or stimulate a search for new technologies to overcome problems that earlier social activities bequeath to subsequent generations.

Each generation's perception of its own landscape, and of the earlier technologies and labor that made it what it is, forms part of the systems of knowledge necessary for social production and reproduction. Accepting the landscape as given constitutes a particular view of history that may restrict the potential for new technological solutions, including those that might retard or reverse the ecological damage that results from earlier environmental modifications. Envisioning the landscape as malleable, in contrast, constitutes a particular view of the future that may open the potential for new technologies. Depending on the ways society uses these technologies, some might damage the environment; others might conserve or protect it; and all will affect the ways that access to resources and control over what is produced from them are controlled and distributed. Attitudes and beliefs about the ways the landscape was formed in the past, and about the ways it could be changed in the future, thus critically shape a population's beliefs about the interaction of society and nature, about the distribution of power and property between groups and classes, and about what can be done to change social or natural arrangements.

In this book, I consider how changing sets of social knowledge, organization, and technology mediated the relation between a particular community and its landscape. I examine the ways that earlier populations constructed a complex set of environmental modifications—irrigation ditches and associated terraces in Huanoquite, a pueblo high in the Peruvian Andes. I then consider how a series of exogenous political and economic changes—the Spanish conquest and colonization, the rise of hacienda control over land and labor, the growing dedication to commercial agriculture, and the agrarian reform of the 1970s—led to neglect and abandonment of the technically most complex ditches and to a loss of the knowledge that had generated them. The collapse of these ditches enhanced erosion already set in train by a simultaneous de-intensification of agriculture and de-diversification of crops. The combination of imposed exogenous power, environmental degradation, and increased orientation to commercial agriculture weakened and distorted the social systems of meaning and authority that constituted the community's relation to landscape—and that organized and regulated its stewardship of the ditches. The maintenance and use of the ditches that continue to function are now fraught with conflicts between different factions of the community with distinct rights in land, different economic interests, and different forms of social organization. I therefore examine how these social and material changes affect the ways that contemporary Huanoquiteños view and explain, use and abuse, maintain and neglect the ditches, and the landscape, that they received as gifts from their ancestors and that continue to support their economy.

I have three purposes in writing this book. First, I wish to consider the general problem of how environmental modifications are transmitted from earlier to later generations; how later generations use, maintain, abuse, forget, or neglect the technologies these environmental modifications required; and how technologies designed to make the environment more productive often end up degrading or destroying it. Second, I believe that the location, construction, maintenance, and use of ditches reveal a great deal about tensions between collective and individual needs, interests, and production strategies, how these have changed over time, and how they are affected by different class relations, systems of authority, and distribution of resources. Third, I believe mountain irrigation systems—and the extensive skill, insight, and labor that imagining, building, and maintaining them requires—can teach us a great deal about the evolving interactions between society and nature, and perhaps about ways we humans can manage the social side of that interaction better to reduce its damage to the natural side.

The research reported here emerged from the intersection of my interest in social perceptions of the environment with Linda Seligmann's larger study of the attitudes, beliefs, knowledge, and practices of the individuals and groups who had attempted to transform Huanaquite's agriculture. I found that my attempts to understand the ditches and their relation to topography led me to questions about the same phenomena she understood better than I, from myths about social origins and history to social and political organization to distribution of and conflict over resources. Our joint research, based on a year of field work in 1984 (Bunker, 1985; Bunker and Seligmann, 1986; Seligmann and Bunker, 1994; Seligmann, 1995) focused on the loss of technology catalyzed by the 1970s land reform. More recently, after many spiritual and mental journeys back to Huanoquite, I realized that understanding fully the loss of technology required that I understand as well the origins, innovations, and implementation of that technology, as well as what this would reveal about that society's relations to nature; about how technology and cosmology interacted with and participated in social relations of dominance, and in social relations of collaboration; and about how these interactions changed as new systems of dominance were imposed from without this society.

My questions about how these ditches were built and operated led me to look at landscape and at the ways that different generations of Huanoquiteños have taken it into account and used it. To do this, I complement data I gathered through interviews and observations of current human activities with data I gained from systematic observations of as well as bodily movement across the physical environment in which current and past human activities have occurred. I then try to relate what I can learn of the eco-cosmologies that evolved with these activities to my own physical practice of experiencing—moving through, working in, and looking at—that environment.

For this reason, much of the book is devoted to analyzing what the original ditch builders would have had to learn, understand, and do to fit ditches to their landscape. I attempt to replicate their experience of imagining and building their irrigation system by constructing a "thick" description of the ditches and of the landscape they cross. I have kept this description as concrete and specific as possible, as an exercise in observation for myself and for readers patient enough to follow it. Just as I had to look at the same topographic features several times, and from several different places, to see their relations to other parts of the landscape, so parts of this description are deliberately repetitive as I combine multiple perspectives on the same natural and social phenomena. I use maps, rather than photos, to illustrate these relations.

I found that ditches and terraces, and the slopes they cross or flatten, are elusive photographic subjects. Most of these photos are not as striking or beautiful as they could be; their aim is not to impress with the beauty and drama of the landscape, but to instruct. Often I took photos to show important detail when a shift of angle would have resulted in a technically better, but less informative, picture. Even so, these images are at best imperfect substitutes for direct looking. Distances and depths are foreshortened, and pictures cannot communicate the experience of standing in different places and turning the head and body, combining sight and movement in forming one's own vision of places.

Irrigation systems provide dramatic cases of received environmental modifications that can be used long after the technologies and knowledge required to build them have been lost. Ditches must be cleaned and repaired, but they can last for years after the original work of tracing and building them across continuous slopes has been done. They change the characteristics of both slopes and soils, through erosion, through soil depositions, and through the kinds of depletion or accumulation of organic and mineral matter that result from the forms of cultivation they make possible.

Because the ditches last so long, and because their collateral effects are usually slow and cumulative, irrigation systems form part of an accepted, taken for granted environment. Later generations may lose the technologies necessary to design and build the ditches; they may also devise new technologies to exploit or to correct the collateral effects of irrigation.

A successful irrigation system brings both new land under cultivation and allows agricultural intensification, and increased commerce and consumption, for larger populations. Inhabitants of irrigated lands learn and come to depend on cultivation practices and crops appropriate for wet fields. Ditches do not last forever, though; the slow processes of erosion to which they contribute may undermine their own bases, or may impoverish the lands they irrigate, so that the ditches may have to be rebuilt or located in other areas. The loss of the knowledge, technologies, and social organization required to trace, build, and maintain ditches and terraces may, therefore, have drastic consequences—for society and for nature—if the system deteriorates too far.

These technologies, and their environmental consequences, emerge from the interactions of society with nature. These interactions, in turn, occur where socially constructed eco-technologies meet a naturally constituted landscape. To understand these interactions, we need to attend to the myths, rituals, beliefs, and stories that people tell about their technologies, and about the places where they use them. Therefore, before we proceed with our analysis of how the Huanoquiteños modified their particular landscape, and how these modifications affected their environment and their society, we need to consider what we know, or can know, more generally about the interaction of nature and society, and about how technology mediates and changes that interaction.

How Technology Mediates between Society and Nature

Anthropology has sustained more coherent attention to the interaction between society and environment than either sociology or geography. In numerous discrete analyses, anthropologists have given us extensive inventories of relations between culture and ecology in hunting, in agriculture, and particularly in irrigation. Attempts to theorize these studies, however, have generated intense controversies between proponents of extreme environmental determination of culture, on the one hand, and advocates of radical social construction of nature on the other (see, for example, collections edited by Descola and Pálsson, 1996, or by Whitehead, 2003). In carrying on these polarizing, and often ahistorical, debates, anthropologists have tended to ignore the contributions of other disciplines to the study of how technology molds an ongoing, reciprocal, and reiterated interaction in which culture and environment shape each other in ways that cumulatively transform both. In other words, they have neglected exemplary evidence that both society and nature can be conceived as having autonomous agency within a dynamic interaction of each responding to, and being transformed by, the ongoing changes in the other.

Marx's dictum that technology mediates between society and nature has framed analyses rooted in notions of nature's and society's mutual catalysis by thinkers as diverse as economic geographer David Harvey (1999), historical geographers D.W. Meinig (1993) and Peter Hugill (1994), and economic historian William N. Parker (1991). All of these scholars are exceptionally attuned to the ways that technology must conform simultaneously to human purpose and to the material and spatial organization of the physical environment.

Parker, in particular, considers two anthropological implications of this principle. He posits that: 1) those individuals who can imagine effective technological innovations must have some special awareness of the natural environment as it affects the material processes that the innovations would engage, and 2) those individuals who can implement new technologies must have some special ability to communicate their technological and environmental insights to their fellows, and to mobilize them around the requirements of the new technology. His formulation of technological innovation concatenates: a) the materio-spatial attributes of particular local eco-systems, b) the personal and the cultural attributes of innovative individuals and societies, c) a clear recognition that culture is simultaneously learned by individuals and evolved by collectivities, d) a realization that individuals vary in their learning of and contribution to their evolving culture, and e) historical changes in the larger social, political, and economic environment, including conquest, migration, subordination, exploitation, domination, and expansion or contraction of territorial or imperial control.

Parker only refers to these implications obliquely, as a possible explanation of how diverse immigrant cultures first mediated, and were then transformed by, individual and collective responses to locally specific material challenges and opportunities on the American frontier. He does, though, cast this conception of society-nature interaction within a historical dynamic driven by the accumulating changes these responses bring about in the physical and in the social environments. This enables him to consider abstractly how the possible lived experiences of the environment change as new technology and different kinds of social organization evolve.

He does not, though, directly examine the processes by which individuals generate new ideas about how to conform the physical environment to human purposes. Nor does he address the anthropologically more exciting question of the ways that different individuals—and the different cultures in which they live—learn, know, and communicate with and about their particular environment. Thus he does not explore the spiritual and practical engagement with animals, plants, earth, water, wind, and sky that appear in Rival's (1996), Arhem's (1996), or Whitehead's (2003) explorations of other peoples' technologies, cosmologies, and landscapes, nor the pre-linguistic ways of knowing and the extra-human communication that enable these engagements.

I believe, though, that Parker's perspective on technology, culture, and nature can be extended to incorporate these pre-linguistic ways of understanding and practicing nature. Earlier, Rappaport (1979) and Bateson (1972) provided us with similar prescriptions of the mental conditions required for full material understanding of relations between nature and society. The lesson that Hornborg (1996:47) derives from these authors, that "the linear structure of purposive, problem-solving consciousness is incapable of grasping the circular connectedness of living systems," could extend and deepen Parker's perspectives on how sensitivity to the environment can mold and support effective technological innovation. Understanding could also be promoted by Arhem's (1996:201) finding, in those same authors' works, the possibility of "world-wide" indigenous representations of nature-society relations and "their integration into totemic, animic, and more complex eco-cosmological models as cultural codifications of deep ecological insights, developed during millenia of intimate practical interaction with the environment."

I work to incorporate these insights into an extension of Parker's perspective. I recount my search into the interaction between particular individuals' special sensitivity to the environment on the one hand and the imagination, skill, social organization, and community labor needed to translate this insight into working technologies on the other. I sought answers to the questions Parker didn't ask by examining and comparing the construction, organization, functions, and maintenance of three irrigation systems in the community of Huanoquite, about 20 km. south of Cuzco. Huanoquite's location, high in extremely rugged mountain terrains of the Peruvian Andes, made the construction and maintenance of these ditches particularly challenging. To learn how Huanoquiteños understand the relations of the ditches they built to the landscape they live in, I participated in the physical and social practices of ditch construction and maintenance and listened closely to the multiple local forms of communication—instrumental, organizational, distributive, juridical, ritual, mythological, historical, prescriptive, and proscriptive—about the social, the natural, and the supernatural aspects of their ditches.

As research and analysis progressed, combining participation in physical and social practice with close attention to the interaction of the social, the natural, and the supernatural stories as recounted in myths and stories underscored the historical fact that Huanoquite had experienced a sequence of radically different forms of community organization, each with different systems of power, authority, and collective mobilization. These had progressed from local self-determination through increasingly centralized, standardized, and absolute exogenous control, from the indirect rule of the Incas to the sporadic, but often deadly visits of civil and clerical colonial authorities, to the progressively more oppressive though still fragmented power of the *hacendados*, through the growing presence of a liberal state committed to increasingly urban centered, bureaucratically standardized, and highly legalistic social and economic programs aimed at increasing its own power and legitimacy. The capacity for fully engaged, ecologically attuned collective effort tended to diminish as authority became more standardized and centralized in national jurisdictions and programs.[1]

---

1 This finding directly contradicts Wittfogel's (1957) association of hydraulic systems with despotic bureaucratic regimes. It also counters a subtler assumption in many studies of Andean irrigation, that complex irrigation systems indicate coercive control over

Each of these power systems had distinct relations of property in and control over land and labor, and by extension over water. Analyzing physical and social process within each system revealed very different relations between persons, classes, and communities. Within each system, these social relations were played out in and affected community relations to and understandings of the environment. As a result, each system adapted to the environment, but in different ways and with different degrees of efficacy.

The historically changing relation within the continuing context of interdependence between systems of authority and society's relation to nature brings out more sharply Parker's understanding that different cultures are differently open, and conducive, to individual sensitivity to the environment, to technological innovations that emerge from that sensitivity, and to collective mobilization for the transformation of physical landscape and of social practice that new technologies entail. As we examine the physical and the social evidence about the construction of ditches in this chapter, we will take into account as well what we know about the different systems of dominance that may have affected both their construction and their maintenance.

The Huanoquite Water Works

All three of the ditch systems I studied come from the same source. The oldest, highest, and most complex one, that irrigated the largest area, was either pre-Incaic or Incaic. It started just below the place where the Qewar river gushed up to 450 liters of water per second powerfully out from under a rocky outcropping. This extraordinary font was one of the meaningful places in Huanoquiteños' explanations of their own history and of their oppression. Some of their most politically, economically, and culturally charged stories and myths about spirits of earth, water, and animals revolved around this particular site. The ditch itself, which had run high enough to irrigate various distant communities in entirely different drainage basins, had been abandoned due to washouts. Many of the stories explained its collapse and recounted the resulting fragmentation of social links between and within the various communities it had served. There were tentative plans for its restoration in the 1980s, but stories about an ancient curse on anyone that attempted to restore it generated fear and opposition.

Another ditch, still in use and probably no more than two or three hundred years old, starts considerably lower down the narrow valley from the same water source. It once provided water to local haciendas, to the village of Huanoquite, and to a number of small hamlets nearby. The haciendas, divided in the 1970s between their rural workers, persist as the telluric bases of agricultural cooperatives, so some of the divisions of interest and power between haciendas and villagers continue in tensions between cooperatives and villagers in present day organization to maintain the ditches. The stories about this ditch revolve primarily around conflicts between different users, the difficulties of mobilizing *faenas*

---

labor. This assumption persists despite broad agreement that Andean political systems remained decentralized and egalitarian even under Incaic rule.

to clean and repair it, and its actual deterioration. They are thus quite distinct in their form and significance from the stories about the oldest ditch. A third ditch, smaller and of indeterminate age, but that I believe was built earlier than any of the major ditches, was originally built to serve a village far below in the Ccorca valley, but two large landowners who kept their lands through the 1970s land reform now take most of its water. The stories about this ditch also involve conflicts over tenure, use, and maintenance.

The construction, use, maintenance, and collapse of these ditches has transformed the landscape and the social organization of the communities that the ditches serve. In that sense, the social history of the landscape and of the communities are written in the ditch systems. The ditches are also, though, a direct manifestation of both a) the unchanging principles and universal physical rules of weight, gravity, conservation, inertia, and flow that govern the natural world, and b) the material processes these rules regulate. Landscape-changing social behaviors join these unchanging natural principles to the far more contingent and malleable stories that human agency engenders within the constraints of physical law, no matter how these laws are conceived.

Dueling Ethnographic Perspectives on Nature, Society, Landscape, and Ditches

Irrigation systems provide wonderful cases for exploring the most fundamentally creative human responses to the environment, and the clearest examples of the simultaneous, autonomously interacting, agencies of both society and nature. Irrigation is arguably the oldest form of technology after the domestication of plants and the origination of agriculture. The peculiar physical, environmental, and social constraints on hydraulic systems—most notably ways that gravity combines with the fugacity and weight of water in their functioning and in their propensity to collapse—force their conformity to the specific topographies and geologies of the places they are built. They also require governance systems that mobilize all users to maintain the system and regulate them in the distribution of its water.

These direct environmental constraints on technology and on social organization have spawned various attempts to identify social and political patterns common to all irrigation systems. The best known, or perhaps most notorious, are Karl Wittfogel's (1957) attempts to explain what he termed "oriental despotism" (essentially his own gloss of Stalinist communism) and Marx's attempts to explain what he called "the Asiatic mode of production" in terms of the governance requirements of irrigation systems. These over-ambitious attempts to theorize broadly, and singly, environmental and technological shaping of society, politics, and economy by the physical properties of irrigation systems have quite justifiably faded into oblivion, or at least intellectual history. Less well known is Wittfogel's (1928) earlier, empirically better grounded and ideologically less tendentious, analysis of how the topographies of different Chinese valleys shaped not just locally specific irrigation technology but also locally specific social organization.

Ecological anthropologists in the 1960s, though abandoning the search for general patterns and universalizing explanations, preserved Wittfogel's earlier close attention to specific topographies to elaborate what they called the Steward-Wittfogel hypothesis (Harris, 1977, chapter 13 reviews this literature and the surrounding debate). This was in fact more a general approach to analyzing the ways that environments affected social organization than a formal hypothesis, but it did spawn a number of descriptively rich studies of technology and social organization in a variety of irrigation systems. Though these studies do not specifically address the question that concerns me, they do provide ample data about and insights into the tight interaction between natural and social systems that irrigation systems require, and I have used their insights and ideas to guide my own field experience and data analysis.

Theoretically more relevant to my search is Basso's (1996) recent work on the ethnography of places and the construction of place-histories as a culturally significant interaction between particularly sensitive individuals and the communities in which they live. Individual imaginings mold what is known of history and of environment into stories of what transpired in particular places. Basso (1996) focuses on the extent to which place, and place-histories, record and teach information and instruction about social relations and about correct action, but the stories he records also reveal a huge sensitivity to the particular material, hydrological, and topographic features of specific places and to how those features affected social behavior. I used a very similar technique to his methods of evoking Western Apache stories about place *in situ* in my analysis of what stories mean and how they communicate meaning about the ditches, and about particular places along them.

Building and operating mountain ditch systems depends on social organization to implement, maintain, and operate complex technologies that only work if they are precisely fitted to the specific opportunities and challenges that their particular environment presents. Answering Parker's unasked questions required that I understand both the physical environment and the ways that local communities know and communicate about it. I worked to achieve this understanding by participating in and observing the physical experience and perception of people involved in ditch use and maintenance. I then evoked and listened to their stories about ditch building, as much as possible after or during work or observation with them at the particular places the stories referred to.

I used the physical experience of working on, walking along, or simply watching the ditches to elicit explanations, stories, legends, and rules about the places along their trajectory—and about the problems encountered and solutions found in each place—from their local users. In these conversations, I attended as well to the strength, clarity, and detail of individuals' knowledge and understanding, and to how these differences in personal knowledge corresponded to their apparent authority and reputation in matters concerning hydrologic maintenance, decisions, and divisions. When possible, I had these conversations at the site itself, so that I was experiencing directly the same environmental reality as the person telling me stories or answering my questions, and could use concrete physical referents for my communications.

I realize that culture, cosmology, and language all shape how we humans experience the physical environment. I also realize that culture may so profoundly shape different ways of knowing that communication across them—especially about the highly abstract cosmologies regarding what Western systems refer to as nature—can only be partial, heuristic, and tentative (compare Descola, 1996). Physically experiencing and talking about the environment in the way I did, though, convinces me that there are patterns and regularities in material and spatial processes that all humans experience physically and spiritually, and of which knowledge includes non-linguistic and extra-cultural sensation and understanding. I found that these physical and spiritual experiences translate more easily across language and culture than do the social and intellectual processes that constitute core data for most ethnographic endeavor. These conversations also convinced me that some individuals were far more attuned to the ways the ditch had to be fitted into and repaired within its landscape than were others, that these individuals were able to communicate their insights to others, and that my identification of such individuals fitted with local users' assessments of the same personal attributes.

Considering the two observations together led me to consider 1) whether pre-linguistic or meta-cultural sensitivities to the dynamics of air, earth, water, and fire, unevenly distributed within a community's members, might be an important and continuing source of belief and cosmology, and 2) whether and how belief and cosmology respond both to the local environment and to changes wrought in it by exogenous forces—social and natural—that change the environment, and the community, at historical moments or in long historical time.

Finally, I elicited knowledgeable Huanoquiteños' cooperation and comments on conventional contour maps on which I was tracing the ditches. After I completed the maps, I had long, detailed conversations about them with the dozen individuals who seemed most aware of how the ditches, the landscape, and the community interacted. I believe these activities helped me to understand both the individual insights into the environment that underlie the knowledge required to build and maintain the ditches and the communication between particularly insightful individuals and the community whose labor must be mobilized and coordinated for the ditches to work.

By asking the same physical—material, spatial, geological, and topographic—questions of the same landscapes and in the same specific places within them that not only the contemporary users of the ditches, but also their original designers and builders, had to take into account while communicating with and about earth and water, and by sharing this experience of inquiry with contemporary users of the ditches and story-tellers about them, I was able to reduce the culturally ingrained tendency of all observers, no matter how sensitive or sympathetic, to look at other people's cosmologies as exotic solutions to problems their own cultures have already figured out. At the same time, though, the sharing of the physical experience about which I was conversing protected me from making the radical social-constructionist interpretations of, e.g., Eduardo Viveiros de Castro (1996) or Descola (1994, 1996).

These scholars are so committed to the idea that different peoples construct not only their own cosmologies, but also the very environments that those belief systems predict and explain, that they neglect Marglin's (1990:15) distinction between organic propositions, "the truth of which depends on the belief of agents," and atomic propositions, "the truth of which is independent of these beliefs." I agree with Hornborg (1996:57) that Marglin too sharply divides the social and the natural when he concludes that "propositions about the world of things and plants are atomic, while many if not all propositions about the world of human beings, the world of social relationships, are organic," but the fundamental distinction between the two types of truth is essential to working out the complex articulation of, and interaction between, society and nature.

There is no question that the Amazonian landscapes that de Castro (1996) and Descola (1996) observed have been significantly modified—intentionally and unintentionally—by human activities, and that these activities and effects form part of the stories that are incorporated and transformed into cosmology, but to claim that such human impact and interpretation invalidate the kind of appreciation of fit between cosmology and belief that Reichel-Dolmatoff (1976) theorizes is to construct an ultimately and arrogantly anthropocentric cosmology of our own. Their critiques somehow miss that Reichel-Dolmatoff based his attribution to the Tucano and the Desana of system-theoretically realistic beliefs about nature on their shamans' ability to perceive and understand humanly catalyzed changes in the environment. He was particularly impressed by the shamans' capacity to operate simultaneously on, and to translate between, abstractly invariable principles incorporated into their cosmology, their manifestation in ongoing material processes in their own environment, and the stories they constructed to direct their human fellows away from activities and places before they started to excessively reduce the energy that plants and animals there could contribute to their well-being.

A refusal to acknowledge the unseeable but immutable powers and laws that regulate the material processes that humans can perceive and manipulate leads directly to a failure to appreciate how different eco-cosmologies construct extraordinarily diverse accounts of physical invariables. This failure, in turn, denies to the ethnographer an entire quadrant of what makes the study of human society so fascinating. An ethnographic willingness, instead, to see myth and "ritual as vehicles of knowledge about the unchanging, transcendent order behind appearances" (Arhem, 1996:202) vastly expands our insights into and understanding of both society and nature. Acknowledging that there are natural systems of rules and laws autonomous from human agency, and that human belief systems are constrained to conform to their visible manifestations, in no way detracts from the agency, autonomy, perspicacity, and creativity of different cultures' extraordinarily distinct cosmologies. Rather, the human genius and creativity evident in such highly varied social construction justifies deep awe in human participation in and engagement with natural systems, however they are conceived.

Different cultures, and perhaps even different individuals within particular cultures, evolve multiple and diverse ways and degrees of knowing, understanding, and communicating the abstract physical laws and regularities that govern, and make comprehensible and predictable, the material world (see, e.g., Descola's (1996) and Arhem's (1996) comparisons of various Amazonian cosmologies with each other and with world-wide systems). Some cosmologies elaborate complexly specified relations between the unseen and the apparent, while others seem less concerned with specifying the nature of general or constant, though unseen, forces and rules. Some cosmologies conceive of society as distinct from, others as intimately articulated with, nature.

Descola's (1996:98) conclusion, though, that the western concept of nature is so vague as to be meaningless seems to emerge from his neglect to distinguish between nature as physical laws, unchanging and unseeable, and nature as the dynamic, constantly changing, and highly visible material manifestation of these laws as they configure differently in different spatio-material combinations. This is a particularly strange lapse, as his own ethnographic work (1994) is organized around the forests, animals, gardens, and rivers that the Jivaroans can construct and use precisely because they have constructed their own understandings of these physical laws in ways adequate to their own purposes and customs.

Nature can be conceived of as unchanging, unseen laws impervious to human action, though humans construct the myths, rituals, metaphors, and models that represent these laws. Human action, though, can catalyze dramatic, sometimes irreversible reorganization of nature conceived of as the varied and variable material forms that these laws regulate. The same myths and rituals account for, recount, and regulate these historic actions and effects. Hornborg and Arhem make clear in their reworking of Rappaport and Bateson's ideas that humans construct mythic and ritual representations of nature's laws—with varying degrees of adequacy and insight—in order to know and communicate how they should act to catalyze the changes they desire, or to impede the ones that they hope to avoid, in nature's material manifestations. All four of these authors clearly distinguish between nature as invariable constant laws and nature as a highly vulnerable material system; failure to keep this distinction clear makes the concept of nature seem vague, confusing, or even meaningless.

Anthropology can strengthen its growing ability to approach others' means and ways of knowledge on their own terms without resort to anthropocentricism. We can gain insights into others' ways of knowing by experiencing with them what that knowing is about. However much culture mediates our experience of nature and of landscape, talking about particular places along the ditches while sharing others' physical experience of these places achieved far greater ease of mutual understanding far more rapidly than I have experienced when asking about purely social or cultural phenomena.

Recording and examining the sensations, ideas, and communications that this practice made possible convinced me that physical and material process, or more specifically the ways that earth, water, fire, and air communicate with humans, do have meta-cultural impacts on human sensations and perceptions. As social scientists, we should be ready to accept these communications and the knowledge that can be gained from them. Thus, we can take seriously, and incorporate into our theories, the invariant laws of physics, and the nearly universal rules of chemistry and biology, even as we acknowledge that different languages, cosmologies, and ways of knowing will perceive and represent them differently. Indeed, our acknowledgment of these variations should make clear and present to us how tentative and partial are all human understandings of invariant law and rules that we can never experience directly. Once we understand this, we can incorporate into our ethnographic theories the ways that these invariant rules and laws govern the geological, hydrological, topographic, climatological, floral, and faunal processes that we can experience directly.

Over a century ago, C.S. Peirce (1877) observed that humans devise and modify their beliefs so that actions based on them will achieve their intended results. There are multiple ways of believing and knowing, but there is also an autonomous reality, however we humans represent it, to which our belief systems must conform in order to function for us. We can believe that human social organization is devised to adjust the material manifestations of this autonomous reality as much as possible to human purpose without for a moment implying that our culture's cosmology—or its idea of human purpose—is the superior standard by which to judge others. We know that social process—and the cosmologies, knowledge systems, organization, norms, stories, and human relations that support and embellish it—are far less governed by laws and rules than are material processes, but to the extent that we humans must adapt to and subsist from material processes, our social organization must conform to them sufficiently to achieve our intentions.

The Mutual Catalysis of Belief, Practice, and Landscape

A clear implication of Peirce's pragmatic assessment of belief is that we will work harder to construct beliefs about those things in nature we wish to change, and that we do this work to make those beliefs consonant both with the intentions for, and with the material upon which, we act. In this sense, societies that construct their environment in ways that require coherent connections between interdependent places within and across an extensive landscape (i.e., ditches, dikes, canals, roads, or railroads) would tend to elaborate and articulate precise, and internally coherent, representations of how nature as immutable unseen law corresponds to and acts upon nature as particular configurations of material processes that, in turn, constitute the topography and hydrology of particular places.

Because their builders must take into precise account not just the separate places along a ditch, but the way those places relate to each other through the autonomous reality of gravity, fluidity, soil composition, and slope, topographically complex irrigation systems provide rich insights into how belief systems, or eco-cosmologies, interact with the natural world. Reichel-Dolmatoff (1976), and more recently Basso (1996) and Abram (1996), have described place-sensitive cosmologies based on particular individuals' special abilities to communicate with and about the physical environments of the particular places their communities inhabit. These materially and morally rich ways of seeing and knowing are not easily accessible to western modes of analysis and thought, but all three authors manage to discover, and then to communicate, insights into specific non-western modes of consciousness and perception. Their work shows that these insights can be gained: first, by close attention to the physical environments that mold them and then by close observation of the processes in which the knowledge that both comes from them and constitutes them is gathered and communicated.

Reichel-Dolmatoff looked at Desana cosmology as an environmentally inspired systems theory. His approach, though brilliant, was hampered by his marveling at how these people long ago achieved an analytic sophistication that foreshadowed recent achievements of western science. Basso's endeavors to enter into Apache wisdom on its own terms constitutes a major advance over Reichel-Dolmatoff's earlier pioneering work, but Basso loses one of Reichel-Dolmatoff's strengths in the process. Basso is so concerned to go beyond the human ecologists' unidimensional attention to the imperial force of the environment over social organization that he does not adequately explain how individuals construct place histories by combining what they know from socially transmitted historical stories about what happened in particular places with what they know from more general principles about how these events must have occurred.

Collectively and over time, people discover general physical principles, by searching for emerging patterns and similarities in their direct communication with the environments of many different places, about the opportunities, challenges, shelters, dangers, fertility, and water that each place offers (see Rappaport, 1968; Basso, 1996; Arhem, 1996). They progressively codify these patterns in myth, ritual, belief, and theory that explain how these general abstract patterns regulate specific material and energetic processes. They then use these codified patterns to contextualize the eventful histories of their society, and of their landscape. The stories about social and natural agency must take into account how general abstract and unseen principles configure differently in the different combinations of material, energy, and spirit that shape particular places and the particular historical events that occur in them.

While the forms, scope, elaboration, and complexity of different individual's and different culture's eco-cosmologies vary broadly, my field work convinced me that the individuals who could integrate these different levels of reality and ways of knowing were the most knowledgeable about the relations of ditches to landscape in Huanoquite. They were also the most able to mobilize and direct community efforts to maintain the ditches and to use them properly. These individuals used "natural" knowledge to set a frame within which stories of human and natural agency could be explained, even though they remained unpredictable (compare Hornborg's 1996 explanation of contextual thought). I believe that such ability to move across and between these different kinds of reality and knowing is probably an essential component of all sustainable society-nature interactions.

By organizing my research as a site-specific inquiry and sharing of information with local experts in the maintenance and building of ditches, both as contemporary practice and as recounted in myths, legends, histories, and stories that enforce normative proscriptions and prescriptions on ditch-related behavior and obligations, I attempted to combine the culturological respect for the other that helps Basso to understand Apache thought and language on its own terms with the ecological awe that helped Reichel-Dolmatoff to see how the particular environmental visions and understandings of adept shamans were so obviously right and useful guides to how that environment should be used that right-behaving members of their community would obey their instructions.

Whitehead (2003:59-60) advances these ideas further, pointing out that landscape, as the material manifestation of "the relations between humans and the environment," as well as the ritual and mythic practices that shape it, changes over time. Thus, the landscape is "more than the natural physical environment and refers also to the immanent presence of the biota and topography, as well as the dynamic activities of humans, fauna, flora, and the spirit world. Historical consciousness is also present in a landscape, reflected in the recounting and memory of the way that those interrelationships have developed over time." Whitehead, like Basso and Reichel-Dolmatoff, can theorize the social and historical construction of landscapes, and of the social understandings of the environments in which they are constructed, without claiming that nature itself is socially constructed as do Viveiros de Castro and Descola.

Descola eloquently describes Jivaroan society's concepts of and relations to forests, rivers, and gardens, but he restricts his problematique exclusively to the social when he writes that "the social practice of nature hinges at one and the same time on the idea a society has of itself, the idea it has of its material environment, and the idea it has of its intervention in the environment (1994:3)."

I propose to complement the partial truth of this mentalist formulation with a materio-spatial perspective. I believe that articulating the complex interactions between place, landscape, and immutable physical laws with the more contingent histories of human thought, intention, agency, accident and error will allow me to unpackage cosmologies into 1) beliefs about constant or invariant dynamics in nature that can only be experienced abstractly and indirectly and that can only be communicated metaphorically, 2) understandings of the more directly perceived processes that these dynamics configure in specific times and places, and 3) the narrative stories about directly experienced and perceived events that human agency and interpretation can mold within the constraints of these processes.

Irrigation systems provide instructive cases for analyzing the interdependence between a) a community's sense of place and time as manifest in the stories and explanations that constitute its members' understandings of the environment, b) its rules about how to treat that environment, and c) the special knowledge or sensitivity of individual members who can mediate and communicate between the environment and the community. Ditches irreducibly manifest the crucially interdependent but still autonomous agencies of nature and of society.

In nature, each of the infinitely divisible sequence of places along a ditch is absolutely crucial to the performance of the ditch downstream. The relative position of each sequential place along the ditch determines both its trajectory through and the spatial extent of the landscape that it creates. Ditches are physically more "path dependent" even than other landscape modifications that connect places within a large scale in a logical system of sequences; roads, for example, can avoid obstacles by moving downhill and then back uphill, but once a ditch loses altitude, it also loses upland portions of its potential landscape that cannot be recovered. Thus, natural principles directly shape the created landscape. At the same time, the trajectory of the ditch, and its construction, stability, and duration in each of these naturally linked places also 1) obeys social purpose, 2) takes into account the socially determined division of space into property and use—both individual and collective, 3) depends on technology that is learned, transmitted, and implemented socially, and 4) has direct consequences for the human carrying capacity and commercial potential of the area it irrigates. Expanded population and production in the irrigated landscape will occasion further historical changes, both material and social.

Ditches must work as systems within topographic, hydrological, and environmental constraints, but they only work as systems if each local place along the ditch is treated in terms of its specific properties—slope, soil composition, materials available for construction of retaining walls, etc. Within each landscape, there may be multiple possible trajectories for ditches that would irrigate multiple possible combinations of the landscape. Similarly, the most mysterious, difficult, challenging, or dangerous places along the ditch will have specific stories associated with them. There are multiple possible stories, multiple possible cosmologies that might encompass them, but all of the stories will fit into a coherent cosmological system that explains, prescribes, and prohibits actions and relations in ways compatible with social intention and subsistence, and survival. These invariably interdependent dynamics make irrigation systems clear cases of mutual catalysis between natural and social agency. Culture fits landscape not because the environment determines eco-cosmology, but because culture, as a social elaboration of individually experienced physical and spiritual events, emerges in interaction with the landscape that the society exploits for subsistence, pleasure, and profit.

We can thus conceive of ditches as both engendering and depending on precisely configured, naturally constrained, technologically complex, socially dense relationships between places and people within articulated landscapes. This would enable me to extend Basso's, Reichel-Dolmatoff's, Descola's, and Whitehead's work by specifically linking the kinds of knowledge they analyze to technology and to social organization. It would also provide additional perspectives to earlier studies of how colonial conquest and political change led to the loss of indigenous technologies essential to both human subsistence and to the integrity of the environment, such as those of Ross (1978), Hemming (1978), Alden (1976), and Sweet (1974) on how slave raids and European appropriation of land for imported cultivars and herds destroyed Amazonian adaptations to their environments and the various stories in Palmer and Parsons (1977) of how imperially imposed systems of extraction, agriculture, and trade occasioned the deterioration of linked social and natural systems in Southern Africa.

Huanoquite: Ditches and Places in a Rugged Landscape

Huanoquite, a pueblo of over 900 people, is located at 3,383 meters above sea level (masl.) 21 km. by foot and 65 by truck south of Cuzco. Like most of the Apurimac valley into which its rivers flow, the terrain around Huanoquite is rugged, varied, and steep, especially in comparison with the Cuzco and the Vilcanota-Urubamba valleys (Escobar, 1977). The sharp changes in elevation provide ecological and climatogical zones appropriate for a wide range of crops—potatoes, corn, barley, wheat, favas, broad bean (*Ulica Faba L.*), tarwi and other leguminous cereals, and forage for cattle, sheep, and horses. In nearby communities closer to the floor of the Apurimac, citrus, *tunas*, and other fruits are cultivated, and in others higher up, llamas are raised for both meat and wool.

The changes in elevation, however, make transport difficult and dangerous, and limit cultivation to less steep slopes and shelves separated by both distance and height. Marked wet and dry seasons require careful scheduling of cultivation to avoid root rot in one period and drought in the other. Annual frosts further restrict the agricultural calendar. Irrigation ditches, terraces, and drainage canals can relax some of these ecological constraints, but the mountainous terrain makes both the construction and the use of ditches even more difficult than around Cuzco and in the Sacred Valley.

Huanoquite is endowed with great quantities of fine ground water. In comparison with other communities in the Apurimac valley, it is blessed with exceptional extensions of flat and fertile land. These two advantages, however, are offset by a single, centrally important, topographic fact; a long high mountain, and a deep narrow gorge down its south side, separate the Qewar, the area's most voluminous river, and the only one which sustains its flow through the dry season, from the best and flattest agricultural lands. Overcoming this obstacle required great conceptual leaps, necessary even to imagine irrigating under such adverse conditions in the first place; extremely precise knowledge of topography in order to chose the trajectories capable of bringing water around these obstacles; sophisticated construction technologies capable of taking these trajectories across steep loose slopes, and a huge and disciplined labor force to incorporate these technologies into the ditches and to maintain them against the force of water rushing down hillsides to break their walls and undermine their bases.

Most of the oldest ditches were abandoned long ago, but the Huanoquiteños continue to live with the environmental effects of their collapse. The more recent ditches continue to function, but the social and political conditions under which they were built, and the economic organization they served, have changed radically. Until the 1970s, Huanoquite's agriculture, and its politics, were dominated by the owners of a few medium-size haciendas. These were actually quite small in comparison with the enormous holdings in the north of Peru; even the largest included little over 1,000 hectares, but they did control most of the cultivable land and absorbed much of the available labor as long ago as 1689 (Villanueva, 1982). By the 1950s, growing urban demand for wheat to bake and barley to brew, enhanced by the expanding spatial extent of roads and by the increasing material efficiencies and decreasing cost of truck transport, intensified the production demands on labor and left less time for stewardship of land and water.

The tightly centralized control of land and labor under the haciendas facilitated the mobilization of labor for the construction and maintenance of ditches and the organized distribution of the water in them. Physical evidence and legends about the earlier construction and maintenance of the older, topographically more challenging, systems built under or before the far less coercive indirect rule of the Inca state suggest, contrary to Wittfogel's notions of environmentally determined despotism, that local communities guided by knowledge-based local authority and solidarity can construct and maintain technically and organizationally complex landscapes.

The agrarian reform of the 1970s adjudicated much of Huanoquite's land to the community of Maska, which makes up over two thirds of the pueblo's population. It also transformed a single large hacienda into a cooperative, Tihuicte, owned by the former *feudatarios* of the hacienda. A few of the old haciendas remained as they were, divided between multiple heirs but still constituting relatively large private holdings.

The reform changed patterns of landownership, of labor relations, of production, and of consumption. The land adjudicated to the community is worked harder and fallowed less frequently than it was under the haciendas, while the remaining large landowners have less access to labor and some of them now work their land less than in the past. In order to maintain the ditches and distribute their water, the community of Maska and the Tihuicte cooperative must now reach some agreement with each other and then organize the labor of members who are no longer subject to the much greater powers and discipline of the haciendas. Thus, even though the ditches continue to flow in the same channels, the social, political, and economic relations of the society that uses them have changed enormously, affecting the ways the ditches are perceived, used, and maintained. This study focuses on the ways that the ditches were constructed and what this construction can tell me about past and present perceptions and uses of the environment, but the analysis in it both requires and facilitates understanding of how the control over land and labor, and the ways these are combined in agricultural production, have changed over time.

Various analysts of Andean society recognize the importance of irrigations systems, but their work focuses largely on ritual and religious aspects of ditch maintenance (see Fonseca's 1984:61 comments on Arguedas, 1964; Isbell, 1974; Ossio, 1976; and Sherbondy, 1982, in this regard; see also Zuidema, 1964.) Even more scholars follow Murra's (1975) lead in considering the ways that social organization in the Andes reflects the dramatic, highly varied, and difficult terrain there. Surprisingly, however, there has been little consideration of how irrigation systems were built and work, and what the cognitive and physical requirements for these processes can tell us about social organization and knowledge of the people who construct and use them. Mayer and Fonseca (1979) and Fonseca (1984) describe the social control and the continuity of irrigation systems within the environmental constraints and changing purposes of different production zones, and Montoya, et al. (1979) show how an ideology of equal access to water masks differences in power and property control which apportion highly unequal benefits to ditch users. Kosok (1965) and Moseley (1974) discuss the technologies of irrigation, but their work concerns the very different, less difficult, northern coastal areas. Mitchell (1976) and Netherly (1984) provide important assessments of the impact of social organization on the construction, maintenance, and management of irrigation systems. Even these authors, however, do not see that our understanding of the labor in and the conflicts over the individual use of a collective resource can be greatly enriched by analysis of the mechanics of ditch construction and use as part of social knowledge and social organization.

Ditches: Technology and Landscape

Conventional or classic irrigation systems typically involve the relatively simple task of keeping some of a river's water above the lowest point on its valley floor. As the river, and its valley, flow downward, the river seeks, and carves, the lowest course. Thus, the valley's sides lose water. The ditch builders' task is to find the minimal downward slope across the valley walls, maximizing the irrigated area between the ditch above and the river below. If, as is often the case, the surrounding topography slopes in the same direction as the river flows, the ditches may create the illusion of actually climbing the valley walls. This is especially the case when ditches are used to take the water from the valley below up to flatter areas through which the river has cut its course.

One such case occurs in northern New Mexico, in the Rio Hondo valley near Taos. Standing at the head of the valley, the sharp contrast between the river's course down its valley and the ditch's minimal descent along the valley wall, whose top slopes west at about the same angle as the valley below, make it almost impossible to see that the ditch emerges onto the surrounding plain at a point lower than that at which the ditch leaves the river. The men who built this ditch had to be able to visualize this, and must have had a precise conception of the land around them, in order to imagine the ditch at all.

Ditches in Chacon, New Mexico, provide an even more impressive example of local comprehension of complex topographical relations. Here, ditch builders made water "climb" the walls of a high mountain valley by taking advantage of the southward slope of the entire hill system. They dug a ditch that took water from the south-flowing river at as easterly an angle as was possible while sustaining a constant, but minimal, downward flow. The water they thus diverted reaches a low point in the mountain wall, and comes cascading down the other side over a long waterfall and into a narrow canyon which leads it to the Chacon valley, where warmer, wider, flatter, and less rocky soils create a highly productive agriculture. There is no single point from which it is possible to see all of the topographical features which the ditch builders had to take into account (see also Rivera, 1998), so deep and intimate knowledge of the environment, and the ability to communicate with and mobilize their neighbors, was essential to the success of their complex and ambitious project.

The conception and visualization of topography necessary to imagine and then construct these mountain ditches are impressive, but the difficulty of this task pales in comparison with the requirements for ditch building in Huanoquite. The two New Mexico ditches described above are complex, but the water flows generally in a single direction, and at most crosses from one valley into another. In Huanoquite, the slopes which must be traversed are much steeper. The flat areas which can be irrigated are not separated from water sources by a single ridge between a higher and a lower valley, as in Chacon, but are in most cases separated by one or more entire valleys, some of which descend into narrow, steep, and rocky gorges. The slopes which must be utilized do not always follow the same directions as the rivers; in some cases the ditches nearly double back in the direction from which they have come. In order to cross entire valleys, the ditches do not simply follow and take advantage of general slope, but have to go against it, running up, rather than down, a valley. Also, because the valleys in Huanoquite are sharper and shorter, each one provides less water than the rivers in the gentler New Mexico mountains. Huanoquite ditch builders had to join the waters of several rivers or valleys into a single ditch in order to irrigate sufficiently.

Their extraordinary achievement in imagining, designing, and building these ditches was affirmed for me by the intense effort I had to make just to follow where they flowed and to understand how they worked. I thought I already understood about ditches and mountains. I had worked cleaning ditches and irrigating from them in the Sangre de Cristo mountains of New Mexico, and I had worked there as a surveyor of wilderness areas and of public lands, first for the U.S. Geodetic Survey and then for the Bureau of Land Management. Even so, it took many long days and a lot of long, steep walks even to begin to figure out how the ditches ran.

One of the challenges, and satisfactions, of this work was that, in the absence of informants who could tell me directly about the kinds of knowledge and skill needed to design and build a ditch, I had to learn to read the landscape itself. Ditches do not provide the kinds of remains and artifacts that archaeologists use to date human occupations and activities; as one archaeologist friend said, ditches at best can be considered as veins between sites, and with luck one can date them by dating associated sites.

Even if such dating were possible in Huanoquite it would not tell me much about how the ditches were built and used. I could, of course, deduce that building the ditches intensified agricultural production, and I could estimate the population sizes which the extent and intensity of agriculture might suggest. I could also see some of the effects of irrigation. Ditches wash out, and must be repaired, or if the damage is too great, they must be relocated or even abandoned. Irrigation often aggravates erosion; sometimes I found terraces designed to prevent this result; sometimes the erosion patterns above and below ditches were so distinct that I could get some idea of the effect of the ditches, and even imagine how the deterioration of land under irrigation reduced production potential. Once I learned to recognize these erosion patterns, it became easier to trace the course of ditches that erosion and cultivation caused to break or to be abandoned.

Many of the topographical features that determine the use and location of fields in Huanoquite were created, deliberately or accidentally, by previous agricultural uses. I cannot date these with any precision, but they participate in processes that continue today, so my understanding of the ditches and terraces, what they can tell me about the past, and what we can learn from present day Huanoquiteños visions of their land and of the ways that their forefathers modified and used it becomes part of our understanding of what changes present and future uses of this environment may bring about, and what Huanoquite's agricultural options will be.

Learning about and reading from the landscape shaped my data and my analysis of the ditches. I also had to take into account what I could learn about past and present patterns of land ownership and the power relations which correspond to it, about changes in cultivation strategies, about market choices and calculations, and about the options and constraints that climate and elevation create for the different agricultural uses of the environment. In this process, I combined archival and interview data with participant observation of present agricultural processes.

Learning to Look at the Land

One of the most difficult, and essential, tasks I confronted in my study of irrigation was to achieve a coherent mental image of the mountains, rivers, and valleys in Huanoquite and of the Huanoquiteños' concepts of and names for their own environment. I had to gain a sufficiently precise knowledge of topography to explain to myself where and how the ditches flowed; and I had to understand local knowledge about and categories for geographical features in order to analyze the ways people there manipulate and use water resources despite the obstacles of slope and distance.

I had my first view of Huanoquite at the end of a long, frightening, and thoroughly exasperating trip in the back of a large truck filled with people and cargo. Though Huanoquite lies about 20 km. south-southwest of Cuzco, the narrow dirt truck road makes a wide loop, with many sharp turns and switchbacks over long sharp drops, so I approached the pueblo from the southeast, coming up a steep winding road from the Ccorca valley. When I reached the top, I saw a wide, rolling plain, or pampa, stretching out ahead and to my right, with a range of high craggy mountains rising beyond it, and to my left, a steep mountain. The pueblo was ahead of and below us, nestled in a long narrow trough between mountain and pampa.

The pampa stands in striking contrast to the jagged peaks and long ridges that surround it. As I explored the area, this wide, flat space seemed even more incongruous in a region where steep slopes and high summits are the dominant features of the landscape. Most of the mountains appeared to be of volcanic origin, but the pampa's base, under a thin layer of soil, was of limestone thickly veined with chalk. I did find some volcanic rock, usually small fragments, in some of the washes around the pampa's rim, but sandstone and limestone predominated in the piles of loose rock here as well. I also found fossils and shell remains over extensive areas on top of the pampa. I started to form a clearer idea of how the pampa had come to be, when I walked down to the Ccorca, 320 meters below, and saw that the pampa rested on enormous sandstone strata which had been uplifted to a 65 degree angle, all the way from the valley floor up to just under the pampa itself.

It appears that the sandstone base was formed by sedimentary depositions during a long alluvial period, followed by an enormous uplift of sufficient force and scope to elevate these strata to a nearly vertical position. The entire formation subsequently subsided, allowing a second alluvial episode and the deposition of a thick layer of lime and chalk. Finally, another massive uplift occurred, elevating the limestone layer to its present position, and revealing to view the angular disjuncture between the uplifted sandstone strata and the flat pampa.

As I explored further, I realized that many of what I had at first thought to be volcanic formations were in fact hills and mountains which had been formed as volcanic magma pushed up through sedimentary strata. On the slopes above the pueblo, for example, I found places where erosion had revealed continuous sandstone faces, and others where it was possible to see uplifted sandstone strata side by side with the magma that had displaced it. I found volcanic rocks with sandstone fragments embedded in them, and in the bottoms of gullies and washes and at the lower ends of fields, I often found red, green, and gray chunks of basalt rocks intermingled with limestone, sandstone, and large pieces of chalk, many extensively altered by water and heat. Under the northwest wall of the pampa, I saw the white stains of the salts filtered out by years of water. Further away from the pueblo, I found that some of the major mountains were formed of limestone.

I slowly began to realize as well how much of this apparently wild landscape had been reshaped by human activities. Some of the ditches and terraces were easily visible, but it took time to discover the traces of the long-abandoned ones. In one case, I noticed severe erosion and gullying across a ridge which extended past the southeast end of the pueblo. When I asked, people agreed that the land there was barren, that the soils were loose, unstable, and rocky, but no one could tell me how long the hillside had been so bare, or what had happened there. I had seen what I thought might be the remains of a broken ditch high across a deep gully north of this area, but had dismissed these as an old path when my informants seemed surprised at my questions. Months later, sitting in a field near this gully, I saw what might have been ditch traces on a hill across and down from the gully. This time there was an older man in the group, and he said that yes, there had been a ditch there. It was hard to follow the traces. Only occasional stretches, all short, had left even the faintest marks on these steep, loose slopes. Once I figured out where the ditch had run, I was able to ask older men and women what the area under it had been like when they were young. None of them directly associated the severe erosion with the ditch, but several were able to tell me when the ditch had given way, and that the soils on that ridge had once been much more productive.

I also learned to spot which of the flat areas and shelves were formed, or at least enhanced, by terracing, even when the terrace walls had disappeared and erosion had blunted the edges of the built up ground. Sometimes I found rock piles from the broken walls far below the terraces they had formed, sometimes I could guess from the amount of erosion at the edges of cultivated patches that these areas must have been terraced until fairly recently or they would simply have washed away. Occasionally I could not be sure whether flat places on slopes were natural or man made, but as I learned to ask the right questions, people confirmed my observations often enough to make me more sure of my ability to "read" the landscape. As I learned to look, I realized that many small flat plots were cultivable only because of prior human labor that was no longer performed and whose effects were already disappearing under the forces of use, wind, and water. At the same time, I saw that the endurance of these flat places, including extraordinarily long shelves on the slope directly above the pueblo, continued to enhance agricultural production. I also learned to remember the parts of the landscape that were out of sight, but related, through water flow and human uses and strategies, to the land I was looking at. Slowly, the long walks up and down different valleys and hills, and my separate impressions and images of each distinct geographical area; started to form a single image of the relations between the different directions of slope and flow in different valleys, of the ways that hills divided the various watersheds, and of the elevations of different valley walls and floors. I could then start to imagine the ways that the ditch builders might have visualized their own geography. I could also start to see how different slopes, soils, and rock formations, as well as the geographical relation of different production zones, had influenced both the process of ditch building and the consequences of ditch use.

It took months of walking, talking, and looking, however, before I felt any confidence in my own observations. My first information about the irrigation system came from verbal reports, and I was highly skeptical of what I heard. When I asked where a particular ditch came from, people often told me that it came from a source I already knew was two or three valleys over. Convinced that water could not flow uphill, I doubted it could have crossed these valleys. It was only when I set out to walk the ditches that I discovered that the ditch builders had contrived to follow contour lines around mountains in such a way that the water ran counter to the direction of the rivers and of the valleys they farmed, and thus could cross them.

The ditch builders could only have achieved this by calculating precisely not just the slope of single hillsides, but also the overall slope of different valley systems and their heights in relation to each other. As one valley is seldom visible from another, and as the water sources in all cases are completely out of sight from the ditches' destinations, the relations and potential continuities between the slopes of different valleys and hills is not at all evident, even to someone quite familiar with the area. Indeed, the difficulties which my informants had in answering my questions, or even in understanding why I asked, about how water had been brought from one place to another, strongly suggested to me that the knowledge required to build these ditches had been lost, and that most, if not all, Huanoquiteños took the ditches they depended on for granted—a gift from their ancestors which they had to maintain but could not reproduce. Several times, I found that all of the contradictory explanations I had heard were topographically incorrect or even impossible.

The absence of clear contemporary knowledge of how the ditches were designed and constructed, and of how their builders conceived of, used, and worked around Huanoquite's topography, made my research far more difficult, but also more exciting. Following ditches from their sources, I discovered for myself the incredible obstacles the ditch builders encountered and the ingenious and laborious solutions they devised to overcome them. Figuring out topographic relations, even with the aid of modern maps, air photos, aledades, altimeters, and compasses, made me appreciate the precise and elaborate vision the ditch builders must have had, even though I could only speculate about how they achieved sufficient understanding first to imagine, then to design, and finally to build their ditches along the high narrow valleys where water emerges, then across the steep slopes between them and other valleys, and finally to keep them high enough to reach the fields they wished to irrigate.

As my work progressed, some present-day Huanoquiteños became interested in what I was doing. They helped me find old ditch traces, told me stories about water, and taught me the myriad place names that I needed to know to even begin talking to them about the land.

It was only after I had achieved a fairly clear idea of the complex topographical relations and started to piece together tentative answers about how the builders of the irrigation system could have conceptualized the ditches' course that I fully realized how extraordinary the topography was and what a conceptual achievement it must have been even to imagine the ditches. The pampa, the fertile fields below it, the shelves directly above the pueblo, and a high wide valley further away constitute the great bulk of the best agricultural land. Huanoquite's major water source, the Qewar, is separated from these areas, first, by Chuspi Grande, a major mountain formation that extends out from the mountain which rises above the Qewar, then, on the other side of this formation, by the deep, narrow gorge of another river, the Qolla Wayku. The pampa and the shelves above the pueblo are themselves exceptional features in the extremely mountainous terrain, but unlike other parts of Huanoquite lands, they have no water sources of their own, and could not provide the basis for Huanoquite's highly productive agriculture until they did. The Qewar's valley is narrow and steep, with very little cultivable land on its slopes and floor. The idea of bringing water from this agriculturally poor valley across two imposing natural barriers to the flat and fertile lands further east must have required the combination of a great imagination and precise knowledge. The implementation of this idea greatly changed both the face of the landscape and the agricultural potential of the region.

Because Huanoquite's ditches must traverse steep slopes, many of them on unstable soil which has been eroded and gullied by the hard frequent rains of the wet season, abetted by centuries of plowing and harvesting which leaves the soil bare during at least part of the year, the ditch builders had to construct high retaining walls of rock, earth, and *champa* (blocks of sod dug with *chaquitaqlla*, or footplows, see Gade and Rios, 1972). Retaining walls were most frequently necessary on the steepest and least stable slopes; building them imposed great technical difficulties and large labor requirements. Retaining walls built of large, flat-sided rocks are also found at the points where ditches enter into and then out of hillsides in order to cross gullies and washes or to get across rivers in intervening valleys. These must be built of heavy enough rock positioned so as to reduce the erosive pressure of streams and washes. Retaining walls are sometimes used to get the ditch across sheer rock faces. In other areas, rather than constructing walls to hold the ditch up, the builders contained it with rock and soil walls on its downhill side.

In some places, the ditches cross deep gullies or sheer rock faces on *champa* laid across log bridges. Some of these bridges are supported by upright logs wedged into cracks and holes in the rock face. More exceptionally, the builders carved their ditches right into rock faces. This is most common across limestone bases, but I also saw cuts in andesite and other volcanic rock. The skills, courage, and effort required to perform the huge, heavy, but meticulous task of precisely carving a ditch into a sheer rock face must have been extraordinary. If the ditch is properly cut into the rock, the water actually deepens and strengthens the ditch over time. On the other hand, water can dissolve limestone channels, especially when, as is common in Huanoquite, the limestone is veined with chalk.

The retaining walls and the log and sod bridges were technically elaborate, labor intensive solutions to the problems of getting water across broken and precipitous slopes. They were essential, though, to the ditch builders' fundamental goal, to get the water from its source, across long and difficult distances with minimum loss of altitude to fields where it could be used. The need to maintain elevation came from two different requirements for the ditches. First, in many cases, the higher the ditch remained, the greater the irrigable area under it would be. Second, the ditches often had to avoid obstacles which would have blocked their courses completely, or which would have imposed even greater technical difficulty. Sheer rock walls, large boulders, and extremely unstable soils restricted the builders' options; maintaining maximum elevation avoided some of these obstacles, and provided the option of dropping the ditch slightly around others. The ditches' trajectories therefore had to be calculated very precisely.

I found long stretches where ditches drop less than 1 meter in every 100 of distance, despite the extreme difficulties of the terrain they traverse. In some cases, the ditch builders appear to have found the only possible way around major obstacles by devising routes so circuitous that there is no point from which all of the topographical features, or the trace of the ditch as a whole, can be seen. In one of the ditches, this exact avoidance of obstacles occurs several times over a nearly 8 km. course which the ditch builders managed to keep high enough to irrigate most of the best land in Huanoquite and in separate drainage basins beyond it. The Huanoquiteños cannot explain how the ditch builders managed such precision without the benefit of modern surveying instruments, but the exact course of the ditch and its precise location across a variety of slopes suggest that the builders were able to make highly sophisticated calculations of its necessary and optimal trajectory.

I had no way to talk to the men who designed and built the ditches. Instead, I used instruments and maps which the ditch builders did not have, in order to reconstruct the goals they set themselves, the obstacles they encountered, and the technological solutions they devised. I complemented these with extended physical experience and practice of the ditch system and of the landscape it participates in, and extended social participation in chores, *faenas*, rituals, and conversations about ditch and landscape. This combination of measurement with mathematically precise instruments, direct physical observation, and interviews gave me a clear sense of the technical solutions the builders had used to overcome topographic obstacles. They did not, though, satisfy my deeper curiosity about how the builders had understood their environment so well, and how they had shaped a cosmology to orient their community's relations with that environment. It was only years later, after rethinking the stories that their descendents told me, that I should interpret my own pre-linguistic communication with earth and water and animals to learn what the builder's vision of their own landscape was, and what they must have known and understood in order to conceive of, and then to design and build, these ditches.

Mountains, Rivers and Valleys

Huanoquite's ditches take water from three rivers; the Qolla Wayku, the Qewar, and the much smaller Watanay. The Qolla Wayku starts as a small slow stream from soggy ground just below the Abra de Vilcacunca. It flows northeast toward Huanoquite through a shallow gorge about 2 km. long, dropping from 4,030 masl. to 3,800 masl. Then it flows another 2 km. along the south side of its wide, wedge-shaped upper valley, which narrows from a 1.5 km. long west rim at 3,800 masl. to the point where the Qolla Wayku is joined by four smaller streams at Unuraqina (3,661 masl.), just before it descends, still flowing northeast, into a second gorge, 1.25 km. long, and much narrower and steeper than the first (see map for more detail).

We have already seen that Huanoquite's major water source, the Qewar, is separated from the best agricultural land; first, by a major mountain formation that extends out from the mountain which rises above it, and then, on the other side of this formation, by the deep, narrow gorge of the Qolla Wayku. Now we consider how the Qolla Wayku Valley constituted both a great opportunity, and a huge obstacle, for intensified agricultural production.

The Qolla Wayku breaks out of its gorge 1 kilometer west of, and only 12 meters higher than, the pueblo of Huanoquite. Most of Huanoquite's most important agricultural areas lie higher than the pueblo itself, on the slopes rising south of the pueblo or on the large pampa north of it. The major cultivation areas at elevations lower than the pueblo are separated from it by a rise whose lowest point is 3,418 masl., just southeast of the pueblo. This means that irrigation waters for these areas must pass higher than the bottom of the gorge, a topographical feature which greatly complicated the ditch builders' task. The difficulties they faced were aggravated because the Qolla Wayku alone does not provide enough water for all the land they wanted to irrigate.

The Qolla Wayku's waters come in part from filtration from small springs in its upper valley walls, but flow also depends importantly on run off from rain waters, so it varies considerably from wet to dry seasons, flowing least when it is most needed for irrigation. Even at the height of the rainy season, moreover, its flow at the bottom of the gorge, after most of its tributaries have already joined it, is less (265 liters/second, February, 1984) than the Qewar's (335 liters/second, February, 1984, ranging up to 447 in June and 414 in August, the peak of the dry season). The Qewar is thus much more important for Huanoquite's irrigation. Its course, however, is further away, and the Qolla Wayku's gorge separates it from most of Huanoquite's arable land. Fortunately for Huanoquite, the Qewar's main valley is much higher than the Qolla Wayku's lower gorge, but the builders still had to discover a means to get the Qewar's water across the Qolla Wayku's.

The landscape inspires awe. The Qewar emerges full and powerful from under a high crag, falls through a steep forest, and rushes over a series of small waterfalls before plummeting over an enormous one almost at the end of its course. The Qolla Wayku plunges over the narrow waterfalls through its tight, deep, lower gorge. The ditches that disperse these rivers' waters are no less wonderful, but their marvel is that human beings have imagined and built them across a landscape which is as difficult to live on as it is beautiful to look upon (see Bunker, 1985; Bunker and Seligmann, 1986; and Seligmann and Bunker, 1994 for fuller details of topography and construction of these ditches).

The three systems I examined are so difficult, and so complex, that the possibility of linking the high-mountain source of water to the cultivable plains where it would enhance agricultural productivity would have been far from obvious to the system's original designers and builders. The conception and use of the ditches were fully social, and the difficulty and complexity of fitting an adequate irrigation system to this landscape required three distinctly social conditions: a) the sensitivity to the environment and the creative imaginative leaps required to conceive of the irrigation system, b) the kind of social structure that had to be invented to provide the intense and heavy labor, the high levels of physical skill, and the tightly discipline organization needed to build it and maintain it, and finally, c) the consequences—social and ecological—of successive political and economic conquests, subordinations, revolutions, reforms, and rebellions that undermined, degraded, or obscured the ecological meanings, understandings, and accomplishments that sustained social agency in the construction and maintenance of the landscape.

The Inca regime, the Spanish conquest, the alienation and subjugation of land and labor to *hacendados,* expanding urban demand for barley and wheat for brewing and baking; juridical battles over community status, rights, and land tenure; the broadening access to and diminishing costs of bulk transport of crops to market as roads were built, trucks were introduced, and both improved; and finally the agrarian reforms of the 1970s each changed crop mixes and cultivation intensities, labor and property regimes, individual-community relations, and the complex social organizations in which these social forces participate. All of these historically specific social changes reorganized the relations within society and between society and nature in Huanoquite.

Nature rules the material effects of human actions on the landscape, but humans do make their own histories. Humans cannot change nature's rules, however much variability we create in the ways we conceive them. Within the boundaries of nature's rules, though, and through an adequate understanding of what they decree, humans are able to incorporate some of these rules into technologies that rearrange some of nature's processes and products into technologies with which we can restructure—and if we are careless, destroy—our landscapes.

The fit between ditches and topography was relatively easy to discern, even if physically difficult to measure. By using physical experience of the ditch, and by sharing this experience with my informants to guide my questions to them and my interpretations and understandings of their answers, I was replicating the experience of the original creators of the ditch, though I had the huge advantage of knowing what I was looking for, and when I found it, of being able to follow and observe what they had built with nothing to guide them but their own sense and understanding of water and earth. I could then use what I had learned in my physical examination to frame questions and interpretations to my contemporary informants to which they responded with stories and explanations that I could relate back to my contemporary physical and social experience to illuminate the attitudes and actions of my informants' ancestors.

Using these means to study the ditches and the landscape they crossed gave me a fairly vivid sense of what the people who designed them must have understood to imagine the ditches, and then to organize the huge and heavy volumes of community labor to build them, but they did not help me much to understand what kind of experiences and practice would create the special sensitivities to ecology and landscape that such imagination and authority would require. I had noticed, though, that some adults and children emanated a very different feeling—simultaneously meditative and acutely observant—when they were herding or driving grazing animals. Llama drivers, in particular, seemed to maintain a direct, deeply appreciative, communication with the animals they tended and directed. Twice, while exploring steep upland terrains for traces of the old ditches, I took shelter from a sudden rainstorm under a tree that was also sheltering a herder. In both instances, I realized that the herder was watching both the animals and the ways that the water flowed, not just down the hill, but along the nearly horizontal paths that the grazing animals had marked. I believe that this sensitivity to and long, quiet observation of animals, earth, and water would be the kind of experience likely to engender the sensitivity to the environment that would have been necessary to build the ditches.

As my research progressed, my questions became more technically complex and more topographically specific. They increasingly prompted the men with the greatest knowledge of and authority over the ditches to discussions among themselves and, on some occasions, announcements that they had to go to consult an old man who lived high above the village. It was clear that I was not to accompany them on these trips. I knew well the wild terrain where the old man was said to live, and knew there was no human habitation there, but the men I was talking to always brought back rich stories with implied, and sometimes explicit, answers that were as technically complex and as topographically specific as the questions I had asked. The ditches gave me a base from which to enter somewhat into my informants' ways of knowing, but the stories from an old man whose residence on the mountain I at least could not see reminds me that there are still ways of knowing that I did not reach.

I did, though, find rich oral traditions and participated in rituals about the original construction and contemporary maintenance of the ditches. The oral traditions include explicit prescriptions about proper and productive relations to, stewardship of, and respect for the environment, with a strong sub-theme of bad consequences of violating these prescriptions, both for individuals and for the community as a whole.

The most powerful and consequential of these stories revolved around the snake with golden braids. I first learned of this snake when I was trying to trace the remains of the highest, long-abandoned ditch back to its source. I was told that after the Spaniards arrived, someone from Huanoquite killed the snake. The snake then cursed the community, withdrawing from the ditch back into the hole from which the natural river gushed, filling the ditch with large boulders, emerging from the earth as water boiling out of the mountain at a place down below the mouth of the old ditch, and condemning to death anyone who tried to reopen it. Eventually, I realized that the snake was both the ditch and the water in it, and in this sense articulated the naturally determined course and behavior of the water in the ditch with the ditch's absolute social dependence on continued and persistent human maintenance. I had to learn a great deal about the Huanoquiteños' ways of knowing before I could appreciate what this meant. Eventually, I also saw how the ancients could have learned even more from the snake about water flow than the herders I sat with learned from their sheep and their llamas.

The rest of this book recounts my attempts to replicate and to understand the experience of landscape and the technological and topographic understandings and imaginations that would have enabled ancient Huanoquiteños to build and maintain their complex irrigation system across the extraordinarily difficult terrain on which they cultivated crops. I combined my own kinetic and conceptual experience of the landscape, my participation in and observation of the work of maintaining and using the ditches, my interpretations of stories about their origins and about the animals that protect them, and patient answers to my unending questions in an attempt to experience as well the ancients' communication with and about earth, air, water, and with the animals that mediated and protected this communication.

In the chapters that follow, I combine close observation and thick description of this landscape with contemporary Huanoquiteños' memories and stories to reconstruct how their ancestors' communication with earth and water, and with each other, gave them the powers to modify their own landscape so dramatically. The most vivid, and the most compelling, of these stories recount the assistance and protection of powerful animals during the times when the ditches were built and properly maintained, and of these animals, and the earth spirits with whom they were allied, turning against the pueblo when the Spaniards suppressed local authorities, together with the rituals and the community labor they supervised. Chapter Two considers how the ancients learned enough of the rivers and hills in their environment to design the ditches that would remake their landscape. Chapter Three recounts the building of the ditches themselves. Chapter Four considers other environmental modifications that changed the landscape, including roads and minor ditches, and Chapter Five returns to the issues about nature and society, specifically comparing the Andean with the Amazonian perspective, and considering the geological, the hydrological, and the mythological relations that link the two different geographies.

Chapter Two

Mountains, Rivers, and Valleys

Huanoquite is endowed with a spectacular amount of ground water. It seeps from hillsides, runs out of small springs, and even bursts out of hillsides as full, fast-flowing rivers. People in Huanoquite are proud of their water, "pura y cristalina," and tell many stories about its mysterious sources and its magical powers. Some claim to have heard underground rivers; they tell of hollow places under the earth that carry water from wild sacred sites to tamer places where humans can put it in ditches and use it for agriculture. Many of their numerous place names are water names, and many of the spirits they talk about are water spirits.

Water, and water spirits, are natural phenomena, bound by invisible and invariable laws and rules that make water behave constantly under specifiable conditions. These rules and laws are manifest in regular processes that can be measured and numbered, such as the yearly cycles of rainy and dry seasons, the rates of flow down different slopes of hill, or the rates of erosion that result on certain types of soil at certain angles of slope. Somewhat more abstractly, but still observably, water passes through regular and predictable cycles and forms—rainfall, percolation and filtration into and through the ground; evaporation off of the ground; drainage into, passage under, and emergence from the earth; and aggregation through the dendritic patterns of tributaries into progressively larger rivers and ultimately to the sea, where the cycle of evaporation into clouds driven by wind generates rain over land again.

At a far less abstract, and far less regular, level of reality, water can do great harm, and can provide great benefits, to society and to the environment. Water may come, flow, accumulate, water crops, create mud, erode slopes, cause floods, carry objects away, and drown people and animals in contingent or conjunctural events. The rates and intensities of water movement that determine these contingent events are themselves determined by the confluence of other regular processes—winds, temperature variations, earthquakes, droughts, and inundations, that configure with water (cf. Tilly, 1995), and thus structure its spatially and temporally specific manifestations and history in ways that can be explained, but not predicted.

Interactions of water and land thus create contingent geological and social histories out of the configuration of regular processes. These histories create landscapes. Societies that occupy, subsist from, and transform these landscapes must construct cosmologies that respond to all three of the levels of reality that structure these histories. Different cultures construct different cosmologies to codify and relate the ways that invisible laws shape empirically observable regular cycles and processes, and how these have separate laws or regularities have configured in time and place to construct specific physical landscapes comprised of geological and social histories. Each of these cosmologies, in turn, is composed of systems of belief, cycles of myth, explanations, and narratives that account in some way for these interactions of law, process, and history.

The cosmologies that explain water and its effects must be particularly elaborate in societies that depend on complex irrigation systems. Huanoquiteños' cosmologies are no exception—they are complex and dynamic. They consider, interpret, and recount—in ways that guide future, and explain the results of past, social behaviors—water's relations to land. These relations are reciprocally catalytic: land forms water's direction and rate of flow at the same time that water's flow forms the land in place-specific configurations and temporal concatenations of multiple physically regulated natural processes. Sociohistorically constructed technologies, organization, and practice mediate and shape these reciprocal relations. The effectiveness of these social constructions is determined jointly by the nature of the materials they are designed to manipulate and by the degree to which their designers succeed in conforming to the invisible, abstract, and in that sense ultimately unknown natural rules and laws that govern the processes the designers are aiming to manipulate.

This chapter considers the interaction between these cosmologies, the physical environment, the regularities of water and land, and the processes by which technologies are invented and implemented in a particular physical environment. I first outline the natural relations of land and water that the Huanoquiteños had to deal with, then I examine how these relations shape their cosmology, and finally I consider how their cosmologies, their social practices, and their social relations shaped and were shaped by their beliefs about land and water. To make these connections, we must pay close attention to the physical relations of land and water in Huanoquite, in order in the next chapter to consider how the Huanoquiteños built their ditches, their landscape, and their history around these relations.

Many Huanoquiteños believe that the underground rivers come from Qaranqa, one of a string of lakes, each higher than 4,200 masl., each surrounded by high crags, atop a long wide range. Fewer Huanoquiteños have actually made the trip through the Abra de Vilcacunca, a pass 3,900 masl. about 7 kilometers southwest of Huanoquite, then down to the river that flows through Qenqonay, 2 kilometers north of the Abra, and finally up the steep long slopes to the lake.

That the deep Qenqonay valley divides the earth's surface between Huanoquite and Qaranqa, and that the surface drainage of the basin in which Qaranqa is located empties toward the Apurimac river and away from Huanoquite does not make the underground flow of water toward Huanoquite less credible to those who have been there. On the contrary, the rugged beauty of these lakes and of the stone walls and rock faces around them, the majestic wilderness of the upland plains around them, the abundance of tall grass and large water fowl, and the commanding views over the Apurimac valley affirm the sacred and the practical significance of this place, and its provision of the abundant, marvelously clean water that fills their irrigation ditches.

Our task in this chapter is to understand the relation of rivers to hills and valleys, and then to examine how the Huanoquiteños' ancestors fitted their ditches to the natural landscape in order to construct an environment more conducive to productive agriculture. I try to replicate the sensations, perceptions, and understandings that the ancestral builders of the longest, highest ditches experienced as they learned the natural landscape, as they first imagined where the water could flow to reach their most promising fields, and then as they constructed the ditches with which they would transform both the nature and the society in their landscape.

To do this, I will extrapolate from my own sensations, perceptions, and understanding as I learned the same landscape many centuries later. As I noted in the last chapter, I have the huge advantage of being able to see the ditches they had built. Many, though, of the ancients' physical and sensational experiences—distance, direction, slope, elevation, perspective, and vision, all formed by the landforms that constitute Huanoquite's production zone and its drainage basins—either remain the same or have changed in ways that I believe I can understand and take into account. As did the ancestral ditch designers, I needed to combine visual stimulus with the bodily experience of movement and effort—up, down, and along—to experience the sensation of each place, and then to combine these multiple place-based visual and kinetic perspectives with memory in order to reconstruct the experiences of these places into a hydrologically and tellurically coherent landscape. The larger landscape was too grand, too complex, and too filled with vision-blocking mountains and valleys to enable a single systemic overview. A detailed comprehensive vision constructed from several distinct perspectives was therefore essential—first for the ancestors to assure that the water flowed as intended, and, centuries later, for me trying to understand how they had achieved this outcome.

For them and for me, the flow of water shaped the land and showed us the relations and connections between land masses, and at the same time the ways that these landmasses channeled and directed the water. They, and I, had to attend closely to these reciprocal relations, first in the places we could directly experience at particular moments and places, and then in the broader, remembered context we had to create by joining together in systematic sequence the memories of those places and perspectives we had experienced separately.

To take advantage of the consonance in our physical and sensation experiences of this landscape, and to help me move from my direct apprehension of my own experience to a comprehension of theirs, my description alternates between and tries to integrate my own sensations and experiences and those of the ancestral ditch builders through three distinct, but clearly articulated moments and perspectives. Starting with the Qolla Wayku, which has less water flowing in its course but which drains the agriculturally most important slopes, I follow each river from its source down to the pueblo, then on to the pampa, and finally to its descent into the Ccorca. Then I describe the landforms that bound the rivers from the perspective of this downstream trek. Next, I describe what I, and they, could see of these same water courses and land forms looking from the flatter, more open cultivable areas back up toward the rivers' sources. A wider view is possible from the perspective of these lower, flatter places, but detail of soil, rock, and water are lost in that distance. Finally, I show how the ancestral ditch builders transformed their understanding of the landscape into the design and construction of their ditch system. My descriptions of these different experiences incorporate visual and the kinetic experience into my growing understanding of contemporary Huanoquiteños' stories told about these places, and about the landscape they constitute.

The three rivers that feed Huanoquite's ditches, the Qolla Wayku, the Qewar, and the much smaller Watanay form very different valleys and stand in very different relations to the surrounding hills and mountains. In the following sections, I will describe these relations for each river in turn.

The Qolla Wayku

The Qolla Wayku valley, and the pampa that rises to the east of its lower course, just before it begins its final descent into the Ccorca, constitute the bulk of the agricultural land that has made Huanoquite a center of population and of trade since before the Inca conquest. In that sense, the river defines the core of the pueblo's production zones. After starting as a small slow stream from soggy ground just below the Abra de Vilcacunca, it flows northeast toward Huanoquite through a shallow gorge between Llamapintay and Makapampa. These are the two largest, highest mountains in Huanoquite's production zone. Makapampa forms an extended western wall whose steep eastward slopes down to the Ccorca river from its north and its south ends moderates in its center to form gentler, longer slopes apt for agriculture. Water from these slopes cut what eventually became the Qolla Wayku River and valley. These gentler slopes and scattered wet spots were probably the site of the pueblo's initial move away from its cultivation of the lower, warmer, wetter more sheltered, but much smaller spaces into the more extensive, but more demanding, fields its members needed to use as their population grew.

Llamapintay forms a southern wall for, and extends the cultivable area of, this valley. Similarly to Makapampa, its slope into the valley is far softer than the rest of its flanks. In addition to the extensive gentle flanks and small streams they offer to agriculture as they slope down toward the Qolla Wayku, both mountains are topped by extensive rolling plateaus above 4100 masl., and both offer fields apt for dry farming.

The upper Qolla Wayku drops relatively slowly through a fairly narrow gorge, from 4,030 masl. to 3,800 masl. in about 2 km. At the end of the gorge, it breaks out into a long, wide upland valley, almost 1.5 kilometers across at its western upper end. Makapampa's southern and eastern slopes feed the small streams and increase the river's initially insignificant flow. For the next two kilometers, the Qolla Wayku flows at the base of Llamapintay, even larger and higher than Makapampa, and then along the flanks of the gently sloping ridges, Wankankalle and Cerro San Miguel, that step down to the north from the eastern end of its plateau, dropping slowly toward the slopes above Huanoquite and the rolling pampa below it. San Miguel pushes the river up against the southern base of Chuspi Grande, which is itself an eastward extension of Makapampa north of Llamapintay, on the other side of the river. The river becomes significantly more embedded as it flows between these large and closely positioned hills.

Though all but its western flanks are very steep, Chuspi Grande's summit is flat and fertile. The gentler slopes of San Miguel also provide fertile land. Both mountain's tops and upper slopes thus offer extensive cultivable space lower than the sources of potential irrigation water, but they themselves are quite arid. Their plateaus are only slightly lower than the highest possible water sources, and there are significant territorial obstacles, including steep slopes, between them and these sources.

These two mountains thus combine a particularly vivid combination of opportunity and challenge of the sort that Parker (1991) identified as generating technological innovations in cultures that responded to individual sensitivity to the environment with a willingness to radically transform social practice and labor process. We will see later how the original ditch builders found solutions to the challenges in order to realize the opportunities. We will also see that maintaining this solution over such problematic terrain required knowledge, organization, and legitimate authority that the later Spanish conquest annihilated. The ditches that used to water these areas are now abandoned, and whatever is left of them serves only to concentrate rainfall's erosive powers.

The rolling floor of the middle valley, just above the gorge between San Miguel and Chuspi Grande, used to receive water from the same ditches, so is also without irrigation now, but it does receive some water from local small streams. The middle valley belonged to the community of Maska even before the Agrarian Reform and still provides some of its best agricultural land. Several community members have permanent dwellings here. They combine cultivation of the floor and slopes with easy access to upland pastures. Most of the valley had been in fallow for the seven years before 1984, but was to be cultivated the following year.

Having followed the river through its defining immediate landscape, we can take a wider view of the landforms that feed it water and channel its course. We have already seen that the Qolla Wayku's upper valley is bounded by two major mountain formations—Makapampa-Chuspi Grande on the north and west, and Llamapintay-Wankankalle-San Miguel on the south and east. On the valley's south side, the steep slopes of Llamapintay stretch 5 kilometers almost due east from the Abra de Vilcacunca, before reaching Yachipunku (3,966 masl.), a high trough between Llamapintay and the high, wide, gently sloping plateau on the tops of Wankankalle and San Miguel. The plateau stretches 3 km. northward, before dropping to the gorge it forms with Chuspi Grande on the other side.

Llamapintay's flat, wide (1.2 to 1.7 km.) top consists of grass lands in shallow valleys separated by rocky mounds that rise to almost 4,400 masl. On Llamapintay's south side, away from the Qolla Wayku valley, various communities and *anexos* from lower altitudes use the upper slopes for potato cultivation and for pasturing. These south slopes drop precipitously away from Yachipunku's 4,200 masl. rim to the 2,700 masl. floor of the Ccorca Valley, in less than 3 km. of horizontal distance.

Halfway down this slope, nestled on a series of relatively flat shelves, is the small village of Llaspay, a former hacienda that is now an *anexo* of Huanoquite and whose lands are held and worked cooperatively. The same ditch that irrigated San Miguel's western and northern upper slopes continued on around its eastern slopes, eventually flowing under Wankankalle to the east sloping ridge from Yachipunku, which it crossed and then descended rapidly to Llaspay. Llaspay, though, unlike the other places the old ditch served, still has as well an independent irrigation system, drawn in part from a voluminous flow from one of the three underground rivers that are reputed to come from Qaranqa.

Huanoquiteños say that this river is actually a canal built by their Incaic or pre-Incaic ancestors. They claim that this underground canal passes under a sacred site—Inkaraqay—at about 4270 masl. atop Llamapintay's flat wide summit. At Inkaraqay it is still possible to see remains that Huanoquiteños identify as being part of a water system. These remains include limestone formations, apparently carved, associated with three squares built of separate rocks, evidently the top of tunnels built into the ground. The sides of the tunnels are too straight, and too deep, to invite trespass. My companions showed no interest in exploring these mysterious features of the landscape, and expressed convincing doubt about the safety of my doing so myself.

One tunnel emitted a sound of rushing water as if from far below, but other than this sound, there is no obvious source of water anywhere on this high plateau. The carving in the limestone, though, slopes in ways that would indeed carry water, and show signs of the dissolution that extended contact with water would cause. Inkaraqay lies on an absolutely straight line through two sacred sites, Huanakauri and Yachikauri, that I will describe later, a sacred site on the highest ridge between Huanoquite and Cuzco, and then on to the Incaic temple at Ccori Qancha. I was able to elicit detailed stories about why these sites are sacred, but none about why or how they lined up with each other and with the temple in Cuzco.

Just west of Inkaraqay, still on the top of Llamapintay, are the ruins of Chunkall. Here the remains of ditches and water works are even clearer than at Inkaraqay, though there is no obvious source of water here either. The straightest, steepest route from Inkaraqay to the floor of the Ccorca River passes directly through Llaspay, which is located on a suddenly gentler slope where Llamapintay moderates its precipitous plunge downward just above the Ccorca.

Llaspay's water is said to come from the river that flows under Inkaraqay, and the now abandoned ditch that used to bring the Qewar's water to Llaspay flowed across an easterly ridge down from Llamapintay. The huge plateau atop Llamapintay's huge landmass, though itself totally dry, thus maintains strong historic and sacred connections to water.

Wankankalle, the higher of the ridges extending north from Llamapintay to form the eastern side of the Qolla Wayku's upper valley, forms a long (3.2 km.), wide (2.8 km.) rolling plateau which rises to 3,935 masl. from the trough of Yachipunku trough (3,850 masl.) that separates it from the east end of Llamapintay. In its middle, there is a slight rise to the twin peaks of San Miguel, which then continue to drop more steeply until flattening somewhat to form series of shelves directly above Huanoquite on its east side, and the shelves described above as its western side drops into the Qolla Wayku valley. These shelves, stable, fertile, and, relatively convenient (especially those on the eastern slope), form some of the pueblo's most important agricultural land. During the 1970s, the return of most of this land to the community of Maska was adjudicated through the Agrarian Reform, but the land on the shelves directly above the pueblo remained in private hands.

Wankankalle's western and northwestern sides, which slope down abruptly (from 3,800 to 3,675 masl. in .25 km.) to form the lower wall of the Qolla Wayku's valley, are severely eroded, suggesting extensive cultivation in the past, with irrigation from a ditch that ran along Llamapintay's north slope, fed by a now-abandoned ditch from the Qewar. All but scattered traces of these ditches have been eliminated by cultivation and erosion; it looks as if some of the most severe erosion occurred where the ditch broke and concentrated the downward flow of water.

The severe erosion, and the collapse of the ditch that watered these slopes, is consistent with records that these lands have been intensely cultivated, and intensely contested, since the Incaic regime ended. The entire north face of Llamapintay, the top and west slope of Wankankalle, and the north slopes of San Miguel—some of the most favorable land in Huanoquite's production zone with or without irrigation—comprised a single hacienda owned by Daniel Galdos at the turn of the century. This land was divided among several generations of heirs, some of who sold to members of other families. Alienation of the community's land to *hacendados*, followed by division of the land among multiple family members, could well have led to neglect of the ditches and of their subsequent collapse. Community members now cultivate potatoes, broad beans, and barley, both in individual usufruct and collectively, on the top and flanks of Wankankalle, but this is now all dry farming. They use the slopes of Llamapintay for pasturing sheep and cattle. Four of Galdos's heirs, however, still claim, and partially achieve, privileged access to these lands, even though they formally ceded them to the community during the Agrarian Reform.

Chuspi Grande, a high, round mound about 1 kilometer wide whose summit is also intensely cultivated, forms the northern boundary of the Qolla Wayku's middle gorge. The location and shape of this flat-topped hill, together with the similarity of its soils and rocks, make it appear likely that it was separated from

San Miguel as the Qolla Wayku cut its deep gorge along a fault between them. Chuspi Grande forms the valley's northeast wall. It rises to 3,819 masl. near its west end, and then slopes overall east, toward Huanoquite pueblo, along its gently undulating (3 degrees to 10 degrees) top. Chuspi Grande was previously part of another hacienda, but most of it now belongs to the community. Like the slopes of Wankankalle and San Miguel, Chuspi Grande once was irrigated by ditches that have now collapsed. Individual community members cultivate wheat, barley, and broad beans here, on fine land, watered now only by rainfall.

Pukaqocha, a short land bridge running east-west, joins Chuspi Grande's east end to the long high (up to 3,904 masl.) ridge of Makapampa. These three connected landforms bound the Qolla Wayku valley north and west, separating it from the Qewar's valley. Pukaqocha's ridge to Chuspi Grande joins Makapampa's extensively terraced lower east face at Parapara. The wall that Makapampa forms across the western, and widest, part of the valley nearly closes the valley as it reaches across to Llamapintay. Only the Qolla Wayku's first, shallow gorge separates Makapampa from the west end of Llamapintay. Here too, the river appears to have cut through a single landmass, but here at the top of the valley, there are more small streams than on the steep slopes of Chuspi Grande and San Miguel further down.

Several small springs emerge from the sides of Llamapintay, and a larger spring creates a stream from Pukacancha, a small canyon in Makapampa's wall. There are a number of small springs on the floor of the valley as well. All of these minor water sources, some of which are channeled in short small ditches (less than 15 cm. wide), keep different small sections of the valley moist. Most of these lands, which belong to Maska, now support only rain fed farming. At Unuraqina, the narrow, pointed, bottom end of the valley (3,661 masl.) four smaller streams join the Qolla Wayku. These streams all flow along the bottom end of the valley, and so cannot be used for irrigation in the valley itself. This area as well was once watered by an old, now collapsed, Incaic ditch, but for many years the only irrigation here too has been limited to tiny catchments from the small local streams.

The Qolla Wayku drops only 140 meters in its 2 km. course across this valley. After the other streams—two from under Yachipunku, another from further west under Llamapintay, a fourth that flows down the middle of the valley, and the one along Chuspi Grande's wall—join it at Unuraqina, it flows northeast, now falling much faster through a lower, much steeper, narrower gorge that drops over several small waterfalls and over the large, multi-stepped Catarata del Diablo while falling 300 meters to 3,360 masl. in 1.25 km. The bottom of this narrower, steeper gorge is close to Huanoquite, but the river has by now cut too low for its water to serve the flat agriculture lands on the pampa or on the shelves above the pueblo.

The gorge opens out as Chuspi Grande sweeps away to the north, forming Miska Pata, its long, tall, flat east face. On the other side of the river, San Miguel's slopes turn east here, first over the broken rocky hillsides near the basalt that welled up in the fault between it and Chuspi Grande, and then around to gentler inclines and long wide fertile shelves formed of a highly weathered mix of volcanic and sedimentary rock and soil that form the north-sloping, highly erosible, shelves directly above the pueblo.

Even after it leaves its narrow lower gorge, the river continues to drop rapidly over the increasingly steep slopes down to the deeply embedded Ccorca River. The walls on both sides of it are also steep, 30 degrees near its course and 20 to 25 degrees higher up, except where the Watanay, a small river which runs just below and north of the pueblo, flows into it, breaking the right hand wall and forming a small triangular area of cultivable shelves, some of which have been further leveled by terracing. Below its confluence with the Watanay, the river passes north beneath the even steeper walls, up to 40 degrees, of Ñekes, a ridge which extends north-northeast from the upper, or north end of Chuspi Grande's east face. From the other side, its right bank, rise the walls below the pampa north of the pueblo.

Ñekes ends abruptly, dropping sharply down to a shelf just above the river at Paqpiri, about 1 km. north of the bottom of the gorge. Beyond this point, the Qolla Wayku's left bank slopes more gently up to the wide flat area called Mallma Alta. It then continues north for another 1.5 km. before flowing into the Ccorca, the region's major river and a tributary of the Apurimac, at 3,074 masl. Even here, however, it is deeply embedded, and its waters can be used only to irrigate scattered small ledges along its course.

The Qolla Wayku breaks out of its lower gorge 1 km. west of, and only 12 meters higher than, the pueblo of Huanoquite. Most of Huanoquite's important agricultural areas lie higher than the pueblo itself, on the slopes rising south of the pueblo or on the large pampa north of it. The major cultivation areas at elevations lower than the pueblo are separated from it by a rise whose lowest point is 3,418 masl. just southeast of the pueblo. This means that irrigation waters must pass higher than the bottom of the gorge, a topographical feature which greatly complicated the ditch builders' task. The difficulties they faced were aggravated because the Qolla Wayku alone does not provide enough water for all the land they wanted to irrigate. For irrigation they needed water from the Qewar, a more voluminous river in a much narrower, steeper valley to the north.

The Qewar

The Qolla Wayku's waters come in part from filtration from small springs in its upper valley's limestone walls. This flow depends importantly on runoff from rain waters, so it varies considerably from wet to dry seasons, flowing least when it is most needed for irrigation. Even at the height of the rainy season, moreover, its flow at the bottom of the gorge, after most of its tributaries have already joined it, is less (265 liters/second, February, 1984) than the Qewar's (335 liters/second, February, 1984, ranging up to 447 in June and 414 in August, the peak of the dry season). The Qewar is thus much more important for Huanoquite irrigation. Its course, however, is further away, and the Qolla Wayku's deep gorge and the huge bulk of Chuspi Grande and the ridges that connect it to Makapampa separate it from most of Huanoquite's arable land. Fortunately for Huanoquite, the Qewar's source is very slightly higher than the ridge that connects Chuspi Grande to Makapampa, and its main valley is significantly higher than the floor of the Qolla Wayku's lower gorge. These two details are far from obvious, even to knowledgeable contemporary observers. We will see, though, that they were critical to the possibility of taking the Qewar's water into and beyond the Qolla Wayku's valley, and that knowing and understanding them was essential to the ancients' ability to imagine how to fit ditches to their landscape.

The Qewar rises in a small, shallow draw nestled in the northeast slopes of Makapampa, four kilometers north of where—at the other end of this mountain—the Qolla Wayku's waters start to gather. The river bursts out at 3,777.5 masl., 394 meters above and about 2.95 kilometers west-northwest of the pueblo, from a series of large holes under a 15 meter high limestone crag. It flows through a narrow strip of floor bounded by steep walls for about 300 meters due east from its source, then bends sharply south in an even narrower, rocky bed between another crag on Makapampa on its right and the cone-shaped limestone mass of Wataqasa on its left. After flowing another 400 meters south, to 3,700 masl., it drops steeply, 75 meters down, through a wide grove of trees, to the narrow floor of its main, east flowing, valley, which is bounded by Chuspi Grande's highly eroded north slopes, on one side, and by Wataqasa's south face, whose wooded flank continues all the way down to the narrow valley floor, on the other.

The Qewar flows east through its main valley for about 1.25 kilometers, dropping 90 meters to 3,530 masl. The valley floor briefly widens to 25 meters, but then is funneled into a narrow pass. Here between Wataqasa and a large, sheer sandstone crag Chuspi Grande bends away to the south, to form Miska Pata, its smooth, intensely cultivated east face, whose base drops 170 meters in less than a km. to the bottom of the Qolla Wayku gorge. Wataqasa curves north from this same pass to form its own east face. This face, about 700 meters wide, is steeper than Miska Pata and remains heavily wooded. A long ridge called Martinwayku extends north-northeast from Wataqasa's north end. Under the elbow of mountain and ridge, Mallma Alta, one of the largest, flattest, and most fertile of the private holdings which remain in Huanoquite, slopes down to the lower end of the Qolla Wayku.

After leaving the pass between Chuspi Grande and Wataqasa, the Qewar turns east-northeast and then northeast, dropping sharply in a deep heavily wooded trough between Ñekes, the steep (up to 45 degrees) ridge which extends out from the northeast end of Chuspi Grande, and the slopes of Mallma Alta. It drops 230 meters in 800 of flow, then turns east again, falling another 100 meters over Ñekes' abrupt north end. Here it forms a spectacular three-stream waterfall called Paqpiri. It continues through a narrow canyon that appears to have been extensively terraced, but is no longer cultivated. Finally, it emerges onto flatter ground, but it is still deeply embedded. It crashes over a 5 meter waterfall into the Qolla Wayku at 3,320 masl.

The river's rushing fall is spectacularly dramatic, and subtly violent. Even at its end, where many strong rivers start to moderate onto flatter slopes, the Qewar stays too wild and precipitous to allow direct socialization into irrigation.

The north side of its last steep canyon is formed by a high narrow ridge, actually a brief extension of Ñekes, that separates the last km. of the Qewar's course from a short, but voluminous river, Miski Unuyoq, that rises from a spring on Mallma Alta's slopes and runs parallel to and north of the Qewar. This small area appears to have been an ancient, separate, hydrologically autonomous production zone, perhaps established before the larger irrigation systems were built. There are 7 large terraces here associated with what appears to be a watchtower, probably all pre-Incaic or Incaic. The valley between Miski Unuyoq and Paqpiri also has extensive remains suggestive of terraces and perhaps of stone dwellings. The Miski Unuyoq river joins the Qolla Wayku half a km. downstream from the Qewar's final descent. Even though it runs below the Qewar, the Miski Unuyoq was used to water these terraces. The ancient ditch and terrace builders considered the Qewar's waters too difficult here to tame for irrigation, especially where such steep and narrow slopes leave no room to slow their descent.

To summarize, the two rivers that provide most of Huanoquite's irrigation water offered rich agricultural opportunities but posed major challenges to the pueblo's early inhabitants. The Qolla Wayku starts as a slow trickle 200 meters higher than the Qewar, a little over six km. southwest of the pueblo. It's long slow descent is shaped by the relatively broad valley set between two widely spaced, relatively gently sloped, mountains, both of which offer various small springs that moisten their smooth flanks and then flow slowly into the Qolla Wayku to slowly increase its size to about 240 liters/second. The Qewar bursts full-blown, powerful, and rapid only two km. northwest of the pueblo, then cuts its way rapidly down a rocky, narrow, steep valley. Its valley is dramatically beautiful, but as absolutely inhospitable to agriculture as the Qolla Wayku's valley is welcoming.

Where they flow nearest to the pueblo, the distance between the two rivers narrows from over 4 to about 1 kilometer. The Qewar is 135 meters higher than the Qolla Wayku as they pass their different ends of Miska Pata, but it has dropped almost as far in two kilometers as the Qolla Wayku in seven. The ancestors' task was to slow its vertiginous ascent enough to redirect its flow over or around the steep high mountains that formed its own narrow valley to the more cultivable areas drained by the far less abundant waters of the Qolla Wayku.

The distinct challenges and opportunities these rivers presented to the ancients were enhanced by the mountains that channel these two rivers' flows. These mountains form distinct boundaries around and between Huanoquite's major agricultural areas, create a wide variety of ecologically diverse production zones, and, in the process, channel water into, and at the same time establish formidable barriers between, the two rivers' separate drainage basins. Makapampa's long north-south slope, reaching over 4,200 masl., forms the western boundary for both drainage systems, while Llampintay's long east-west slopes, form the Qolla Wayku's southern boundary, and channel its waters away from the shelves above the pueblo and from the pampa. Wataqasa and the long ridge north-northeast from it form the Qewar's northern boundary. Chuspi Grande, together with Pukaqocha and Parapara, forms an eastward wall through the bounded area that divides the two rivers and their valleys. Like the other mountains the make up the two river valleys, these three linked landmasses offered promising lands to cultivate, but together they also constituted the major, and defining, challenge for the ancients who built the ditches connecting the Qewar's wild-flowing water to the gentler shelves of the Qolla Wayku drainage and then to the pampa.

Wankankalle and San Miguel form a roughly southeast-northwest wall through the agricultural and drainage area, dividing the pueblo and pampa from the Qolla Wayku's upper valley. Across the Qolla Wayku from these mountains, Chuspi Grande's east face (Miska Pata) and the east face of Wataqasa form a north-south wall, but further west of the pueblo. Martinwayku, the ridge which projects from Wataqasa, forms a north wall. If one looks up and west from the pueblo, and then turns one's gaze 45 degrees to the southwest, one sees that San Miguel, Chuspi Grande, Wataqasa, and the ridge form a roughly semi-circular wall sweeping from northwest to south to southeast around the agricultural land visible from the pueblo. The pampa projects east and northeast of the pueblo, following the wide sweep of the Ccorca, whose far more deeply embedded course forms a roughly opposing semicircle from northwest to north to southeast. Thus, the Ccorca valley forms the east, and part of the northeast and southeast, boundaries of the production area and drainage basin that starts high, and far west, at Makapampa; runs between Llamapintay to the south and Wataqasa to the north; is partially split east-west by Chuspi Grande and southeast-northwest by Wankankalle and San Miguel; and finally drops off again, first to the pueblo and the pampa, and then, much lower down the pampa's slopes, to the Ccorca. The builders had to understand the material, topographic, and hydrological configurations of the separate multiple places in these hills and valleys, then see how they fit together as an actual and as potential systems of water flow, and then see how those could connect to the lands apt for agriculture when they imagined and set out to build their ditches.

At present, only the east part of this area—Mallma, Miska Pata, the north face of San Miguel, and the slopes under the pampa's rim—receive significant amounts of irrigation. The Qolla Wayku's upper valley was once irrigated, but it now has reverted to its condition before the ditches were built, with only a few small ditches from springs from its own slopes and floor. The tops of Wankankalle, Wataqasa, and Chuspi Grande—which are extensively cultivated in potatoes, broad beans, wheat, and barley—and of course the even higher *puna* on the tops of Makapampa and Llamapintay, which are used for pasture and some potatoes, receive no irrigation water. These mountain tops and slopes provide spring and runoff water which can be used for irrigation lower down, but the major ditches all flow to the eastern section of this area, to fields around the pueblo, across from it in Mallma, and below the rim of the pampa.

The Watanay

The much smaller Watanay rises on the shelves just above the pueblo, then flows to and through it, and then flows a brief distance through a small, sheltered triangle of intricately terraced small plots before joining the Qolla Wayku. The terraces are old, and in varied states of decay. The Watanay provides small amounts of water to gardens above the pueblo, flows through the pueblo itself, waters terraced plots that appear to have been intensely cultivated for many years, and then meets the ritually, economically, and politically meaningful ancient road to Cuzco. The meeting of roads and rivers in an intensely cultivated area is marked by sacred thrones and sites. Journeys to or from Cuzco begin or end at this sacred place just outside Huanoquite.

The View from the Pampa: Hillsides, Rivers, and Ditches

We have now followed the major rivers down from their sources, and considered their relations to the surrounding landforms and what these would signify for agriculture. This perspective on place-specific relations of land to water provided one critical component of the experience of landscape that the ancients would need to imagine the ditches they would built. They had to complement and integrate this perspective, though, with as broadly envisioned a view of the whole as they could achieve by direct perception, abetted with their memory of the different places as they could place them in this larger view. In this section we will see what this area, and the ditches across it, look like from atop the pampa, which provides the broadest, but still incomplete, view of the landscape at a distance close enough not to lose all detail.

The Qewar and the Qolla Wayku both break out of the mountain wall around the pueblo we have just described; the Qolla Wayku from its gorge 12 meters above and about 1 kilometer west of the pueblo, the Qewar through the end of its upper valley 147 meters above and 1.5 kilometers west-northwest of the pueblo. If we stand on the highest point of the pampa, northeast of the pueblo, we can see the gorge and the pass of these two rivers as breaks in the mountain wall which sweeps around 3 sides, from northeast to south, of the pueblo. We are now looking up the two watersheds from a single point, rather than following them down from their sources as we did in the description above. To the northeast, over the end of the pampa, we look into the Ccorca Valley. Turning counterclockwise, or west, our view follows Martinwayku as it juts out from Wataqasa over Mallma Alta. From here, we can see Mallma Alta's long slope from the foot of this ridge, gradual at first (less than 8 degrees) as it drops, ever steeper (up to 20 degrees) down to Paqpiri and the floor of the Qolla Wayku below its confluence with the Qewar. Looking a little further west, we see the steep ridge of Ñekes breaking Mallma Alta's gentle slope and Paqpiri, the Qewar's long waterfall over the end of the ridge. Beyond Ñekes rises Wataqasa. Continuing to turn, now looking almost due west, we see the Qewar's pass, and beyond it we can glimpse, sideways, the slopes of Chuspi Grande and Wataqasa into its valley. Our view now crosses Miska Pata on Chuspi Grande's east face, and we follow this smooth face across and down until its top half is broken by what appears to have been a landslide, and its bottom half becomes a series of cultivable shelves separated by sharp drops. Turning a little further, we look down into the Qolla Wayku's gorge, but cannot really see up it. It is too narrow, and rock spurs project into it from both sides, creating a zig-zag that blocks our view.

South of the gorge, we see the long north face of San Miguel, dropping steeply at first (up to 40 degrees) over bare rock, loose dry soils, and sharp gullies. As we look southwest across the pueblo, however, it levels off somewhat to form a series of wide (up to 200 meters), flat (5 to 15 degrees) shelves almost 1 km. long, separated by short steep drops (20 to 30 degrees, 5 to 10 meters high). Above these shelves, there are longer, steeper rises, from 3,575 masl. to 3,700 masl., up to the wide flat top of Wankankalle (2 kilometers between 3,700 masl. and 3,875 masl.). The Qolla Wayku's upper valley is out of sight behind these slopes.

As we continue to turn south, toward this ridge, we see a long deep gulley, Huanakauri, running northeast slantwise out of San Miguel's upper slopes. Above and beyond this gulley Wankankalle continues south toward Yachipunku, but San Miguel slopes down a long wide ridge called Tanqarpata. Huanakauri abruptly terminates the flat, fertile shelves above the pueblo. Across it, running down to the east, Tanqarpata is rocky, arid, and deeply eroded. Near the bottom of this slope, the Huanakauri gully intersects with the Watanay gully, which runs straight down hill, nearly due north, before it reaches the trough between the mountain and the pampa, where it turns northwest, becoming a river as it runs below the pueblo. Together, these two gullies seem to follow a major fault line under San Miguel's summit. They also mark important agricultural and topographical boundaries between the dark clay soils, stable and deep, which have formed over the mixture of sandstone and basalt bases on the shelves west of the break and the loose light sandy soils dotted with large volcanic boulders which formed around sedimentary fragments. Tanqarpata is deeply gullied, and dry. In contrast with the heavily cultivated shelves further west, it has only scattered patches of low yielding fields.

Tanqarpata drops from almost 3,800 masl. south of San Miguel's summit to 3,428 masl. southeast of the pueblo. Here, its descending slopes converge with a low rise which connects its base with the pampa rim 300 meters to the northeast. Below this rim the pampa slopes off across a series of large shelves and then more steeply into the Ccorca valley, which swings around in a wide loop from northeast of the pampa to its southeast, from where it continues south to the Apurimac. Above the rim, the pampa rises slowly to the north and east, at slopes which vary from 3 to 8 degrees, to the high point on which we are standing. Further north along this rim, and east of where we stand, nearer to the Ccorca, the pampa rises more quickly, on steeper but still low hills. To the west of us, the pampa's upper flanks are gentler; here, they drop off at 10 to 12 degrees to a series of wide, long shelves between 3,375 and 3,450 masl. Below 3,375 masl., there are narrower shelves, which diminish in width and length as the pampa falls more and more rapidly to the Qolla Wayku. Below the rim, to our north, a wide shallow valley runs to the edge of the pampa, which then drops, precipitously but out of our sight, to the Ccorca. Finally, to the east, the pampa drops over a series of wide, rolling swales and shelves, all the way down to 3000 masl., which provide both good soils, warmer temperatures and more protection from frost.

The multiple variations in soil types, the different stabilities and fertilities these soils produce, the underlying strata of rocks, the slopes and contours they create, their permeability by and retention of water, their angles in relation to the sun, and the relations of their elevation and contour to temperature, wind, and the downward flow of water from higher slopes create a vast array of agricultural challenges and opportunities. These challenges and opportunities constrain and structure human society, but Huanoquite's history clearly demonstrates that different political regimes, with their different cultures and different openness to, generation of, or suppression of individual responses to the environment and the collective transformations of landscape they may catalyze, engender very different interactions with this landscape.

The political regime appears to determine the degree to which the complexity and variability of landscape enhances the divisions between different sectors of the community, or the degree to which the unity of community transcends the variability of the landscape. It appears also to determine social conditions that shape ecologically sensitive individuals' ability to perceive, and to communicate, their physical environment in ways that stimulate new insights into technologies that can alter the landscape in socially useful ways, and then to mobilize their fellows to the intense labor and changes of practice that implementing their technological imagination would entail.

We will see the ways that different political and property regimes affect social interactions with the landscape in the next chapter. As we trace the various ditches, attending as we go to the ways they fit into and have changed the natural landscape we just examined, we will also compare the ways they were imagined, constructed, and maintained under four very distinct regimes. Each of these regimes moved the locus of power and regulation progressively further from local determination. The autonomous pre-Incaic was incorporated into the Incaic empire, which coopted local custom and authority into a highly decentralized coordination based in Cuzco. The intensely Catholic Eurocentrism of the violently repressive combination of clerical and military authority that oversaw the establishment of a colonial state was attenuated somewhat by logistical constraints of time and distance. The Crown's subsequent assignment of local control over land and labor to select individuals gave rise to the autocratic hacienda system, that endured from colonialism through independence. The identity of personal rights over other's land and labor to a particular locality engendered some degree of local responsiveness, but annihilated any autonomous capacity for community determination until a very partial, legally rule bound, set of palliatives in the 1930s allowed for limited recognition of community lands. The more recent attempts to implement a land reform that aimed to equalize access to land and resources, but that tried to implement these goals through highly centralized and standardized bureaucratic and legal procedures mandated and controlled by the national state, were particularly impervious to local variations either of landscape or of community. As we read the ditches and the landscape for local builders' understandings of and responses to the abstract regularities of earth and water, and of how these configured in the material, spatial, and temporal processes of the multiple places they had to connect into the hydraulic system they envisioned, we will also read these ditches and landscape for the changing constraints these different political regimes imposed on the interaction of society and nature.

Chapter Three

Fitting Tame Ditches into a Wild Landscape

As we looked round from our vantage point on the pampa, we saw parts of all of Huanoquite's major ditches, and most of the land that they currently irrigate. One of these ditches flows from the short upper valley of the Qewar, around Wataqasa's south and then east faces, and then along the ridge that projects north from it over Mallma Alta. We saw another ditch leave from the Qewar's main valley, and then swing south across Miska Pata. We could follow its trace above the fields there until it disappears into the Qolla Wayku gorge. It only reappears further south as it crosses between the shelves on the face of San Miguel above the pueblo. Then it disappears again into the gully at Huanakauri and reappears at the base of Tanqarpata, flows along it for a while, and then crosses the low rise between it and the pampa, where it splits into different ditches, one flowing north along the west rim of the pampa, the other flowing northeast and then north along its southeast and east rim. This second ditch waters the many flat shelves that form breaks in the steepening slopes below the pampa's rim. A smaller ditch, higher than the one that reaches the pampa, crosses San Miguel's slopes from above the Qolla Wayku gorge to the large gully at Huanakauri to irrigate the largest, flattest shelf there.

We also saw, through the cleft between Chuspi Grande and Wataqasa, far back on the lower slopes of Makapampa, the traces of two much older ditches, one nearly flat and the other dropping more steeply. They both go out of sight behind Chuspi Grande, but the lower, steeply descending one reappears over the very crown of Miska Pata, runs south along it, then east and then north, still following Chuspi Grande's crown, along the top of the Qolla Wayku gorge.

Both of these ditches have long been abandoned. They watered much larger areas, from much higher trajectories, than any of the present ones do. The ditches still in use, though, do water some relatively high fields; in Miska Pata as high as 3,570 masl. and in Mallma Alta up to 3,550 masl. The same ditch that waters Miska Pata also irrigates fields far below the pampa's east rim, at less than 3,100 masl., and so has a range of over 470 vertical meters, about half a kilometer. It serves many of the different altitudes, and the great variety of soil types, and slopes, that compose Huanoquite's different agricultural zones. Even so, the range of irrigation is only half of the full range of agricultural elevations, which run from below 3,100 to above 4,100 masl. Now that the highest ditches no longer function, far less than half of Huanoquite's total cultivated area receives irrigation water.

We could not, however, from our vantage point on the pampa, see the sources of any of these ditches. The places where the Qolla Wayku and the Qewar break out of the mountain wall around the pampa are already too low for their water to reach Huanoquite, so the ditches start further up, out of sight in these rivers' respective valleys. Nor could we see their final destinations, as these are out of sight below the pampa, where the final descent to the Ccorca river becomes even more precipitous that the rivers' upland flows. Nonetheless, we could see several dramatically striking visual features—for instance, the large number of ditches we could see from a single view-point, the number of times different ditches appeared and disappeared from that view, and the great range of irrigated elevations that could be seen from the same place—attest to the tremendous topographic and hydrological complexity, both of the natural landscape and of the ditches that the local society fitted to it.

Indeed, there is no way to ascertain from where we stand on the pampa which, of the ditch sections that we see appearing and then disappearing connect to each other, nor where or how they do so. In order to overcome the large, steep, and often unstable barriers of hill and valley, the builders had to trace enormously complex, and at first view, counter-intuitive, trajectories. I had to follow each ditch on foot to understand which and how different of the sections that I could see from below actually connected to each other. My first view of these mountains as I entered Huanoquite had puzzled me enormously; I could not imagine why anyone would have built ditches that seemed to hang half way up, and to continue straight to the end of, high mountains that were far too steep to cultivate, let alone to irrigate.

It was only after a great deal of walking that I realized where these apparently hanging ditches went, and how the parts I could see from the pampa fitted together. They actually came down along one side of a mountain (the line I could see from below), then curved around its end (out of my sight), and then went back up against the flow of the valley on the other side (even though continuing to drop, and still out of my sight). When they reached the bottom of the new valley, always at a point lower than, but upstream from, where they entered the valley, the builders had constructed a bridge, or sometimes a trestle, to get the ditch across. Then they would follow back down the valley, but with a minimal drop, so as to break out of it and continue on the flatter ground beyond. Eventually, this ditch would move far enough out of the valley it had just crossed so that it would reappear to me, still looking up from below. The ancient ditch builders devised this complex, labor-intensive, ingenious device wherever a less voluminous river separated a more voluminous river from cultivable land that needed a large volume of irrigation water. The ditches the *hacendados* ordered built centuries later incorporated this same device to overcome the topographic problems that resulted from their lower trajectory.

Combining Views from Different Perspectives

We have now considered the rivers, mountains, and valleys around Huanoquite from the two perspectives I had to combine in order to get a general idea of how the ditches lay and worked—first, down from the rivers' upland sources, and then up as far as I could see from a viewpoint on the pampa. These two perspectives replicate, though in very abbreviated form, the kinds of lived experience, vision, and practice from which the ancestors' would have created their understanding of the landscape in which they would build ditches.

This abbreviated replication, of course, cannot reproduce the long hours, days, months, and years of physical and visual sensation, or of communication with earth, air, and water, that the ditch builders accumulated as they grew up, worked, and aged in this landscape. Having worked at various chores—plowing, planting, irrigating, harvesting, and repairing ditches—on steep slopes side by side with Huanoquiteños, I can attest that the brief year I spent with them as an adult was not sufficient for me to learn to move as comfortably and perceptively in this landscape as they did. I can, though, take this difference in our learning and knowledge into account as I attempt to imagine their physical and kinetic cognition of the landscape. Thus, I can extrapolate from my own physical and visual experience of tracing existing ditches in this complicated, varied, and rugged landscape, to the much deeper vision, sensation, and knowledge the ancestors needed, first to imagine, and then to place them there.

The landscape impressed me powerfully and constantly. The long, broad, gently rolling pampa lies in stark contrast to the steep ascent I made to reach it from the Ccorca. The contrast with the mountains that surround it is even more dramatic. The mountains visible from the pampa rise steadily and steeply to 4,000-meter high plateaus, which then continue to rise more gently from there. I could not see the plateau's gentler rise to their summits from the pampa, though, because my view from below of the long ascent to the edge of the plateau rose at a steeper angle than the slope of the plateau itself. In other words, the steeper lower part of the mountain hides from view the more slowly rising top part of the mountain, so what I saw from the pampa looked like long, straight ridges of multiple distinct, very large, landmasses.

As with the ditches, I got an important overall perspective on the mountain system from a single vantage point open enough to show me how the mountains lie in relation to each other, but fully understanding the topography of each mountain required separate experiences of moving over and along each one's distinctive terrain. Similarly, I had to penetrate the individual valleys in this landscape to appreciate the rivers and the specific places through which they flow. Only after learning of and appreciating their power, danger, and beauty directly could I begin to imagine the knowledge, imagination, and work that the ancients needed to transform such wild and savage, downward swirling, impetuously falling currents into streams tranquil enough to flow laterally across these extraordinarily steep slopes, and then softly enough to irrigate plowed ground without washing it away.

The Qewar emerges from under a high crag, falls through a steep forest, and rushes over a series of small waterfalls before plummeting over an enormous one almost at the end of its course. After growing steadily down its wide upper valley, the Qolla Wayku plunges over the narrow waterfalls through its lower gorge. Both rivers inspire awe. As we will see in the next sections, the ditches that tame and distribute their waters are no less wonderful, but their true marvel is how human beings imagined and built them across a landscape that is as difficult to live on as it is beautiful to look upon.

The Oldest, Highest, Ditches [A and B]

The oldest ditches, now abandoned, are the most impressive—technically, topographically, and socially. They run higher than any of the ones still in use, and traverse even steeper terrain. Two of these old, Incaic or pre-Incaic, ditches are still easily visible, others I saw traces of, and others I was told about. I will first describe the two that are visible from the pampa as they descend the slopes of Makapampa. Each followed a different route to overcome the same obstacle—the barrier that Chuspi Grande and the Qolla Wayku gorge create between the Qewar's constant flow and Huanoquite's cultivable land (see A and B on map).

The builders managed to take the higher of the two ditches not only across a formidably high ridge early in its course, but they then kept it high enough all the way across Huanoquite's entire drainage basin to overcome the high ridge at its other, southern, end. This enabled them to continue the ditch to, and to strengthen their social and economic connections with, Llaspay and other communities in the ecologically distinct production zones along the bottom of the deep Ccorca river valley while also making it possible to irrigate the gently sloping and fertile lands of the upper Qolla Wayku valley.

Incorporating this extensive, varied, and complex terrain into a single irrigation system served by a single powerful river presented the ancients with their greatest conceptual and structural challenges. They first had to imagine and design a constantly dropping trajectory that nonetheless sloped so little that it could cross over hills almost as high as, though several kilometers away from, the place on the river from which they would take water. They then had to build a ditch to follow the trajectory they chose, sustaining a minimal drop for eight kilometers over extraordinarily steep and unstable faces. Doing this enabled them to break out of the natural barriers that huge mountains and deep valleys posed to the goal of sharing their water with communities with different ecologies and climates. In Huanaoquite, at least, Murra's (1975) "vertical archipelago" is rooted in the technology and infrastructure required to share water across vertical and horizontal space.

The ancients' effort to maintain maximum elevation started at the very beginning of the ditch as part of the boldest, technically most complex, and hydrologically most efficient response to the challenge presented by the Makapampa-Paqpiri-Parapara-Chuspi Grande landmass. Rather than taking the circuitous, water-wasting, route down the northern face of Chuspi Grande, and then back around its lower east end, they built the ditch as directly as possible toward the slight dip between Parapara and Chuspi Grande's higher western end. This was a bold, and demanding, enterprise: the dip is over two kilometers away from, across difficult, steep terrain, and only slightly lower than, the ditch's source.

They started the higher ditch (A) immediately below the Qewar's gushing burst out of the rock, then followed on to the right along Makapampa's eastern slope, almost perpendicular to the rapidly falling river valley. To maintain altitude against Makapampa's steep slopes here they had to construct a heavy, 5 meter high, vertical retaining wall of rock and earth along more than half of the ditch's first 50 meters. They built the ditch almost due south over the eastward and then east-southeastward flow of the Qewar for 365 meters, with so little drop that the water in it would appear to be standing still. They had already saved 70 meters over the Qewar's precipitous drop at the end of this distance, just north of the point where the Qewar drops even more steeply over the wooded hillside to its main valley floor.

From here, they followed the hillside as it curves into a wide, shallow draw. The builders obeyed this and other undulations in the terrain closely in order to keep as much elevation as possible. They followed the hill's curve away from the Qewar, west, for 92 meters, then curved south and eventually south-southwest along Parapara's steep (35 degree-42 degree) eastern face, following gullies into the mountain and then back out again. Even here, they had to build extensive rock retaining walls, up to 4 meters in height. Building these walls required a carefully controlled and coordinated series of tasks that would have been difficult under any conditions, and that were made far more arduous by the rugged, sharply dropping, hillside where they had to be performed. These included: moving many hundreds of large, heavy rocks across steep and unstable slopes; anchoring the bottom ones into this slope; mounting the rest above them to over twice a man's height; and leveling the top so as to achieve a gently, steadily, smoothly dropping trough for the water. Long stretches of these walls have fallen, but there are still some sections as much as 20 meters long, often with only short breaks between.

For its first 1.2 km., the ditch crossed steep wooded slopes that show no signs of cultivation. Along the next 400 meters, there are narrow belts flat enough to allow planting for almost half a km. Then, from 1.6 km. to 2 km., extensive terracing on the slopes above and a small amount below widens these cultivable belts. Just beyond this point, however, cultivation and erosion have eliminated all clear traces of the ditch. I used an altimeter and an aledade to follow the ditch's highest possible route at the same slope I had found in the still visible sections of ditch. In other words, I followed the same minimal drop I had already seen along the parts of the ditch I could follow visually. Continuing on this trajectory, I found traces where the ditch builders reached Pukaqocha—the east-west land bridge between Parapara and Chuspi Grande—ran along it briefly to its lowest point, and there dug a 7 meter deep, 3 meter wide cut.

To get the ditch across Pukaqocha through the cut they dug, the builders had to limit its drop to only 22 meters, from 3,777 masl. at the Qewar to 3,655 at Pukaqocha, over 2 km. of flow. The cut has been filled in somewhat by loose dirt since the ditch was abandoned, so it may have been two or three meters deeper than it is at present. The ditch would then have dropped 24 or 25 meters, but no more.

I have seen functioning ditches that drop less than 1 meter in 100, so I know that this rate of drop will sustain sufficient flow for irrigation. The builders of the high ditch across Parapara would have had to calculate this drop precisely, however, in order to be sure of getting across Pukaqocha, before building the ditch. They would have had also to maintain this minimal drop even while building long, high retaining walls on steep and broken slopes. Building such heavy walls of large rock at such a precise angle over such difficult terrain for such a distance without modern instruments or tools can only be imagined to have required a unanimous commitment and willing coordination of intention, knowledge, and effort by designers and many builders working with a common enthusiasm and purpose. Such a strong collective commitment of enthusiasm and purpose, in turn, would have required a deep and consensually legitimated trust in the ditch designers' special knowledge and understanding of earth, rock, and water; and of the hydrological and topographic dynamics that their interaction created.

This knowledge, and its profound ability to mobilize the community's collective efforts, could only have been achieved by deep local practice and experience, and so is incompatible with absolutist or coercive power. The problems of topography, geology, and hydrology, and the complex solutions possible to them, are so specific to particular sites that the idea of imposing an exogenous authority to direct such work seems totally implausible. As we follow and compare the different ditches—abandoned and still functioning—we will see that the ditches built under the coercive power of the haciendas were topographically and technologically less complex, and required less careful and less skillful maintenance than those built under the more egalitarian systems of authority managed either by autonomous local communities or by the Inca's decentralized indirect rule through local custom and cosmology (Zuidema, 1964; Gelles, 2000).

We will see, as we follow the ditch along the next eight kilometers, that the primary motive for this truly formidable achievement was that it enabled them to link their ditch to, and incorporate community labor and agricultural exchange from, the ecologically distinct lands lower in the Ccorca valley. The minimal drop they achieved thus created material and hydrological connections for a "vertical archipelago," that Murra (1975) would attribute to colonizing kin ties, between Llaspay and other villages on the warmer Ccorca valley floor and the higher, colder production zones above Huanoquite. It also maximized the irrigated area in Huanoquite's production zones.

Traces of erosion indicate that water flowing through this cut served to irrigate the other side of Pukaqocha, that is, its south slope. Few conclusive traces of the ditch in the upper end of the Qolla Wayku valley remain, but several considerations affirm its route. I found multiple types of evidence that the main ditch continued around the upper west end of the Qolla Wayku's valley. First, the irrigated area under Pukaqocha was too small to justify the enormous effort of bringing the water that far along such intensely constructed walls. Pukaqocha's south slope is considerably gentler than other slopes into the Qolla Wayku valley, but it is heavily eroded in comparison to all but the steepest slopes down from San Miguel on the other side of the valley. Such erosion indicates a period of irrigation, which could only have come from the high ditch across Parapara. The total area involved, though, is relatively small, so would not alone merit the huge work of building the ditch across Pukaqocha. Second, I did locate a strip of the dense, thorny vegetation, which typically grows in abandoned ditches along the northwest end of the Qolla Wayku valley at elevations that would fit the highest possible trajectory. Finally, after I had traced the ditch and found the long thorny line characteristic of old ditches, one of my most knowledgeable informants described the old ditch as having followed the course I had already calculated. He told me that cultivation had since eliminated almost all of its traces, but that he had seen the traces there when he was a boy.

He said that the ditch had crossed the entire Qolla Wayku valley. The builders would have had an easier time maintaining altitude here, as this wide valley's floor is far less steep than Makapampa's slopes. He told me that the ditch then travelled on the lower slopes below Llamapintay, and from there around and about 100 meters above the base of San Miguel, where this mountain drops to the top of the Qolla Wayku's second gorge. At this point, the ditch turned east to run high above Huanoquite. Here, it could water the shelves that sloped north toward the pueblo.

The extensive erosion on San Miguel's slopes into the Qolla Wayku's valley make it impossible to follow continuous traces there either, but the intermittent traces still visible indicate extensive erosion from irrigation higher up. Beyond San Miguel, though, I found very clear traces of old ditches again. The only way water could have reached these traces was along the route I just described.

How One Sacred Site Embodies Topography, Hydrology, and History

These traces included a long section of ditch dug into a sheer rock face, almost directly above, and slightly beyond, a large rock called Huanakauri, which is situated in a sharp gully whose soils and rocks differ distinctively from those that make up the terrain on both sides of it. Huanakauri itself is sacred to the Huanoquiteños, and figures in their stories about struggles between superhuman ancestors of both genders. Remarkably, though none of the stories I was told mention this topographic feature, Huanakauri lies directly on the straight line that runs from Inkaraqay to Cuzco.

The gully to Huanakauri slices down one side of San Miguel's summit. The deeply embedded Watanay River, which the gully joins halfway down the slope, cuts down the other side. San Miguel is an important *apu*, or sacred mountain, which some Huanoquiteños describe as particularly ferocious, a reflection, perhaps, of the ecological, agricultural, and personal peril of these deep gullies and the rapidly flowing water in them after a rainstorm, and perhaps as well of the extraordinary skill, effort, and risk entailed in building this ditch across the sheer rock face here. The idea that the gullies emerge from and surround this dangerous place emphasizes the topographical and the social significance of the fault line they appear to follow and that so clearly divides good from poor land, as well as of the ongoing dynamic relation here between strong waters and steep, large hills.

Huanakauri also sits on an ancient footpath that has served ecologically and socially important trade through Huanoquite since before the Inca conquest. This path goes to Yachipunku and then to Wankankalle, the wide plateau beyond San Miguel. From there it drops down to Vilcabamba, a community whose much lower elevation allows production of goods with important trade links to Huanoquite.

Standing at Huanakauri and looking north-northeast, a straight sight-line passes first across Yachikauri, the highest point on the pampa below and another important sacred place, and then on to Quipan on the farthest horizon. Quipan is the high point on the old Inca road to Cuzco as it passes over the highest ridge between Huanoquite and Cuzco. Quipan is also an important *apu*. Travellers setting off from Huanoquite on the Inca road make a *tinka*, or offering of food or drink, to Quipan in order to arrive safely and without too much fatigue. On their first journey to Cuzco along the old road, Huanoquiteños believe that they must carry a rock, which they leave on Quipan's summit, carefully placing it with the rocks which other travellers have left there.

After reaching Quipan, if one continues looking along the same straight sightline (that is, looking 180 degrees away from the sight line back Huanakauri from that point), one looks directly at Ccori Qancha in Cuzco. This same line, following back to Huanakauri and then continuing south-southwest, eventually crosses Inkaraqay, the apparently carved rocks high on Llamapintay. Fifty meters south of Huanakauri, just across the gully, there is a cone shaped hill, with terraces on its upper slopes, topped by two circular, nearly concentric terraces which may have been watchpoints, as they look straight up the Miska Wayku gorge, which the Incaic road to Cuzco passes along. Scattered around these soils—which are apparently of sedimentary origin—are numerous rough-surfaced round rocks, all between 2.5 and 3.5 feet in diameter, that appear to be of volcanic origin. These rocks are unlike any others in the area, except for a similar pile of them at the very edge of the Ccorca valley, on a platform high above the river and across from a similar platform on the other side of the valley. They appear to be basalt or some other kind of extruded magma. I asked how the rocks got there, and why there were two similar, but distantly placed, piles of rock so distinct from surrounding geology. I was told that the ancestors had known how to herd rocks, and were in the process of driving the rocks at Huanakauri to the same spot above the Ccorca where the rest were piled in order to built a bridge over the river, to shorten the Inca road that drops deeply into and then climbs steeply out of the Ccorca valley. The arrival of the Spaniards had interrupted this work, the cock had crowed, and the ancestors lost the power to herd rocks.

Huanakauri thus has legendary, as well as topographic, significance. The straight linear configuration of Cuzco with a) Quipan, the highest point between Cusco and Huanoquite, b) Yachikauri, the major *apu* on the pampa, c) Huanakauri and d) the ancient, water related sacred site of Inkaraqay, together with the clear topographical boundary that Huanakauri marks on one of the most important cultivable areas directly above Huanoquite and the far less hospitable slopes beyond it, suggest that these points may have formed an Incaic *ceque* line from Cuzco (see Zuidema, 1964).

Like Inkaraqay and Chunkall, it is unclear whether Huanakauri's various mysterious physical features, all incongruent with the characteristics of the surrounding landscape, result from natural or from social processes and causes. The draw that leads to Huanakauri has far less stable soils that the surrounding land has. These could be the result of erosion from breaks in the ditch, or they could have resulted from hydrothermal eruptions or from landslides catalyzed by tremors.

Huanakauri itself is a single, much larger rock protruding from the soil in the wash below the ditch. I heard stories about a beautiful virgin attempting to flee Cuzco to avoid a magically powerful man who desired her. He caught her as she fled, and either he transformed her, or she transformed herself, into this rock. This story sounded as if it was mythologizing some aspect of the Inca conquest of Huanoquite, or perhaps a human sacrifice by the Incaic conquerors, that was oriented to the divinities that controlled water and made irrigation possible. Such stories would be compatible with the gully the rock is in, and the particularly unstable soil within it, if these resulted from erosion by waters from a break in the old ditch above. Certainly, the feat of cutting the ditch into the sheer rock face above and just beyond Huanakauri, and the horrific consequences of the ditch breaking so close to, and so high above, some of Haunoquite's best fields, could have inspired the kinds of stories I heard, but the coincidence of Huanakauri being on a straight line with other sacred sites also linked to water is stranger, and more wonderful, that I can explain. The coincidence of the *ceque* lines with the Incaic road to Cuzco, however, does suggest richer interpretations of the multiple stories I heard linking sacred site, mountains, water, and travel.

Shared irrigation water, and the ditches and community labor that makes this sharing possible, connect communities in different ecological zones and enable the agricultural surpluses of the different kinds of crops that make trade between them possible and worthwhile. This trade, though, requires transport, and transport is made difficult by the same water that makes trade possible. That is to say, water flows into large, fast, strong rivers that carve deep canyons that impede and complicate trade between different elevations and ecosystems. Fortunately, though, small tributary rivers cut lateral, or perpendicular, slopes down into and then back out of these deep canyons. Most paths and roads do follow water courses; the Incaic road from Huanoquite to Cuzco certainly does.

Water provides the means to make agriculture possible, but it also carves the deep canyons (and makes the mud) that make exchanging agricultural products more difficult. Water also cuts the routes that make it possible to climb into and then out of these canyons, but rushing water makes the river crossing dangerous. Water flows under as well as on the earth's surface and moves through the air as vapor, clouds, and rain. These contradictory relations are powerful, and impinge directly on human and social chances of survival. They all manifest natural processes, but are themselves supernatural. Huanakauri, jutting out under the steep rock face that the ancient builders had to cut the ditch into, is also situated where one can look straight along the waterway that holds the road to Cuzco. It thus embodies the sacred relations between mountains, rivers, ditches, society, agriculture, and trade.

From Huanakauri, the ditch continued east for another 700 meters to cross the ridge into the Llaspay's drainage basin and its ecologically distinctive production zones. At this point, though, we need to return to the Qewar valley, to follow the lower ditch along the alternative route around Chuspi Grande.

The builders started a second ditch (B) across Parapara, branching down from the first ditch 420 meters from its source, just beyond the draw above the Qewar's fall to its main valley. They built this ditch to irrigate Chuspi Grande, whose eastern summit lies lower than its west end, and than Pukaqocha. They could thus run the ditch along, rather than over, Pukaqocha, and thus follow natural contours and a more rapidly dropping trajectory. This allowed them more options for paths that avoided difficult obstacles and unstable ground, so they needed fewer retaining walls.

Chuspi Grande rises to 3,819 masl., towering over the Pukaqocha land bridge at its west end, but the overall slope of its wide, relatively flat top is to the east. This enabled the builders to follow an easier trajectory, with little attempt to maintain elevation. They could drop the ditch enough along Chuspi Grande's north slope to start it only about halfway up the hillside but nonetheless to reach the upper rim, where they had more and flatter ground to irrigate, fairly quickly as they followed its east-tending downward slope. Thus, the ditch drops 30 meters below the high ditch in 1,322 meters of flow, i.e., it drops almost twice as fast as the upper ditch. It runs into the lower north face of Pukaqocha, and doubles back at a 25 degree angle along the west and then north slopes of Chuspi Grande.

Having reached this point, though, the builders were easily able to reduce the ditch's linear drop to maximize the space irrigated. The ditch continues along this rim, from just over to just under 3,700 masl., as it turns across this mountain's east face, high over the fertile soils and gentler slopes of Miska Pata, and then runs west, over the Qolla Wayku's valley, at the southwest end of Chuspi Grande. Here, it flows 30 meters above Unurakina. Now, traces disappear. Some informants affirm that the ditch ended here, but others claim that it crossed the valley and continued back under San Miguel to irrigate the high slopes above Huanoquite. It would have been too low, though, to water all of the fields on San Miguel that show signs of irrigation. It seems far more likely that the only ditch across San Miguel was the high ditch across Pukaqocha, of which I have seen traces, and not this lower ditch around Chuspi Grande, which would not have been very useful and of which I could find no traces. The higher ditch watered a larger area on San Miguel than a lower ditch could have, and the lower, more circuitous trajectory would have lost more water and required more upkeep.

There is little irrigable land below this lower ditch on Parapara, and Chuspi Grande's north face, where it slopes down to the Qewar's valley, is very steep, up to 40 feet in places, with an overall slope of about 30 degrees. There are signs of old terraces toward its steeper, higher, west end. These are mostly above the ditch, where they would have served to conserve scarce water, but some lie below it, where they would have lessened the erosion that irrigation inevitably causes on steep slopes leading quickly to a sharp drop, as they do here. These terraces do not create a large enough cultivable space to justify the work of building the ditch. Rather, the builders' primary destination for this ditch must have been the soft gentle slopes at the east end of Chuspi Grande and fields above Unurakina.

The terraces did not prevent severe erosion, though, on Chuspi Grande's steep north face. The five sharp gullies I referred to earlier all become much more accentuated below the ditch. It seems likely that there was more cultivable land on this slope when the ditch was built than there is now, but this slope was probably never very productive. More significantly, these gullies have broken the ditch, which now concentrates rainwater that would otherwise remain dispersed over a broader portion of the face. These enhanced washes routinely damage a more recently built ditch that parallels the older ditch's trajectory about 100 meters lower.

Miska Pata, on Chuspi Grande's east face presents much gentler slopes (20-30 degrees). Because the foot of this slope drops to Qolla Wayku, 170 meters below the lowest point of the Qewar valley floor, this face also presents a broader, and warmer, area for cultivation than the north slope, which in addition to being higher and steeper faces away from the sun's prevailing inclination. Chuspi Grande's south face, which falls into the Qolla Wayku gorge, is even steeper, and much rockier, than its north face. Except for one wash, which the ditch has to curve into, that was possibly started by erosion from plowing and harvesting, it is hard to imagine that there was ever any significant cultivation or irrigation on this slope, which even animals disdain for pasture. Rather, it seems likely that the ancient builders took the ditch over this unhospitable slope in order to reach Chuspi Grande's south and southwest faces, which border the Qolla Wayku valley, and slope down to the fields on its floor.

The construction of both of these abandoned ditches must have required precise topographic knowledge, great engineering skill, and the coordination of vast amounts of labor. These ditches' destinations, however, would have watered extensive areas of good agricultural land; indeed, the areas under the still visible sections of the lower ditch would have provided significant and favorable cultivation areas. We must therefore consider the reasons for their abandonment, as these may shed light on both past and present agricultural practices, property regimes, and political relations in Huanoquite. Comparing these ditches to those that are still in use will help to illuminate some of these questions, so we will turn now to these ditches, returning later to possible explanations and implications of the abandonment of the higher, older ditches.

The Lower, Still Functioning, Ditches

The two abandoned ditches are spectacular for the engineering of their minimal drops, for the integrated vision required to connect multiple places precisely enough into a coherent and sequential system to maintain these minimal drops, and for the technology and labor that kept the flow of water from the Qewar valley across or around Chuspi Grande and then into the wide, flat valley of the Qolla Wayku. In contrast, all of the ditches still in use flow at lower elevations, so their builders did not need to achieve the minimal drops of the ditches that stayed higher in order to cross over hydrological divides. In most regards, then, the knowledge needed to imagine, the technology required to build, as well as the intense, punctual, careful, complex labor needed to maintain, these ditches were less than those so evident in the high ditch we just followed all the way to Llaspay. Nonetheless, though lower and less demanding, they still impress with their gentle drops across vertiginous slopes, with the skill and labor incorporated into the retaining walls and bridges that carry them across the steepest drops and least stable soils, and with the understanding of mountain elevations, slopes, and contours necessary to imagine and plan them.

Because the ditches followed lower trajectories, the Qolla Wayku, as well as the other minor rivers, was larger and deeper where they crossed them. These crossings presented special technological and maintenance problems. As we saw earlier in this chapter, in order to cross a river that has significantly embedded its valley, the builders contrived to move the ditch's flow against the direction of the rivers to be crossed in order to maintain the elevation necessary to leave the valley on the other side. This required precise technical solutions, which, though different than those of the higher ditch, nonetheless absorbed huge amounts of labor, both in construction and in maintenance.

I was not been able to determine how old these ditches are; there is some evidence that one of them was built around the turn of the last century. I strongly suspect, as I explained in the last chapter, that the one that still waters the area north of the Qewar may well have been the first ditch built in the Huanoquite drainage basin. I will explain in the next section the topographic, demographic, and archaeological evidence that supports this suspicion.

The Ditch to Mallma Alta [C]

The highest of the still functioning ditches, and the one I suspect is the oldest in the region, flows off to the left at the bottom of the Qewar's upper valley, about 700 meters from its source, and just above the point where it plunges over the wooded slope to its main valley. From its beginning at 3,721 masl., the ditch flows south along Wataqasa's base for about 100 meters, dropping steeply to 3,710 masl. During this stretch, much of its channel, including all of the first 40 meters, appears to have been cemented with rock and limestone. The presence of an old kiln, 2 meters across and 5 tall, now overgrown with vines and trees, on the other side of the Qewar makes this seem plausible. However, other parts of the ditch's course have clearly been carved into the limestone base, so it is possible that what appears to be cemented was in fact carved into a base of limestone and rock, and that the water's flow over the years has deepened the channel.

After its first 100 meters, the ditch curves to almost due east as it follows around to Wataqasa's south face. Wataqasa is extremely steep on this side, falling from 3,800 masl. just below its summit to 3,600 masl. on the floor of the Qewar's main valley, i.e., a drop of 200 vertical meters, over a horizontal distance of only 310 meters, a 64.5 percent overall slope. At various points above and below the ditch, the slope is nearly vertical. The ditch clings to this south face for 600 meters, traveling over rock and sod retaining walls that range from 1 to 3 meters in height and which are so narrow (20-30 cm.) that walking along the ditch is very difficult. The ditch is shallow and narrow, its bed is often as little as 25 cm. high and 30 cm. wide, and significant amounts of water spill from it at its low points. There are only two small fields below the ditch on this slope, so most of the spilled water is lost in the dense shrubs and trees that grow there.

The ditch drops more gently here than in its first hundred yards, losing only 27 meters of altitude in 600 of distance, but this fall is still more pronounced than that of the higher abandoned ditch over Pukaqocha. This ditch, however, passes high above the fields it waters, so its rapid drop does not reduce irrigable area, and its faster flow compensates for the small size of the ditch bed itself. The ditch can carry up to 58 liters/second. This is half the volume that the other major ditch—whose bed is over 3 times as large, but which drops considerably more slowly—can carry.

At the end of this south face, 700 meters from its source and 152 meters above the narrow pass through which the Qewar leaves its main valley, the ditch turns sharply north-northwest to follow Wataqasa's east slope for 750 meters, dropping from 3,682 masl. to 3,653 masl. Wataqasa, and the long north-south running hill from which it rises, descend to the more gently sloping, fertile lands of Mallma Alta. Indeed, Ñekes, where the Qewar finally joins the Qolla Wayku at the lower, eastern end of Mallma Baja (the flatter zone where the descending hill levels off a bit below and east of Mallma Alta), borders a nearly flat area of almost a square kilometer.

At the end of Wataqasa's east face, where it joins the ridge, another major outlet drops water through a steep gully, from 3,652 masl. to 3,538 masl. where an auxiliary ditch carries a large stream of water gently northeast another 800 meters along Mallma Alta's upper rim. The auxiliary ditch serves to slow the rush of water enough to irrigate a high, remarkably flat (5-10 degrees), but very small upper area.

The main ditch continues north-northeast along the same contour for another 931 meters, then drops over 3 rocky waterfalls, from 3,652 masl. to 3,566 masl. Here, 2 km. and 400 meters from its source on the Qewar, the long gentle contours of the hill they built the ditch along, which so simplified the builders' task of providing water to the multiple places where flatter, wider, portions of the hillside made productive agriculture possible, start a precipitous drop to the Ccorca river just below. Rainwater flowing down this suddenly far steeper slope cut a deep, northeast-flowing wash here. The builders had to cross the steep, eroded sides of this wash to reach the warmer, wider fields that lay still to the north, near the floor of the Ccorca valley. This goal motivated them to devise the much more complex techniques for carrying their ditches across flowing streams that I described in general terms at the end of the last section. In the following paragraphs, we will see how these techniques fitted the material realities of this specific place.

To maintain elevation, the ancients turned the ditch sharply north to northwest, with minimal drop for 250 meters across the steep rocky point of the ridge high above the Ccorca River. They then turned the ditch west-southwest, into the steep dry wash just where it starts to become a deep, steeply falling, canyon. Shortly beyond the turn they built back into the wash, they had to cross above a narrow pass between the ridge and a sharp crag which projects from it. The ridge here forms a sandstone wall, at about a 70 degree slope.

The builders contrived to get the water over this steep and unstable wash by building retaining walls, and then a log trestle, into the 70 degree sandstone wall that forms the wash's south side. They supported the log trestle, about 5 meters long, on other logs propped into niches and cracks in the sandstone. They laid sod across the trestle, so grass grew into it, fortifying the structure and making the sod less permeable. They slanted the retaining wall and the trestle so that they would take the ditch upstream, that is, dropping slowly uphill (west-southwest) against the more steeply downward (northeast) flow of the wash. When the retaining wall, which was slowly dropping westward, intersected with the bottom of the wash, which flowed northeastward, the ancients built rock supports on both sides and placed a hollowed out log across the wash. On the other side of the wash, they doubled the ditch back, so it flowed downhill, i.e., in the same direction as the wash, but with a minimal drop. When they reached the top of the wash they could turn northwards again.

Previously, another ditch continued east for a kilometer over a long steep (40-50 degrees) sandstone face to reach another high shelf there, but that ditch has washed out. The ditch that drops down to the small village of Wiñay Poqo, 250 meters below on the floor of the Ccorca Valley, through a series of washes and gullies over narrow shelves separated by early vertical steps up to 10 meters tall, still functions. The small fields on these shelves provide the only cultivable land under the ditch beyond Mallma Alta.

The zone that this ditch irrigates is one of the most sheltered, and most easily cultivated, in Huanoquite's drainage basin. It is quite small, stretching south from what is now Wiñay Poqo to the site of the old terraces below Mallma Alta. Small sections of it are naturally watered by multiple small local streams—making dense gardens possible even before irrigation was established. Because it was technically the easiest, I believe that this zone was historically the earliest place where the first inhabitants of what eventually became Huanoquite could learn the technologies needed to grow crops in this particular landscape and then to intensify their production by building ditches. This easier, more sheltered, but smaller zone would have supported an original, sparse, pioneering population until demographic increase required that its members expand their agriculture to the more challenging, far colder and drier, but much larger spaces of the Qolla Wayku valley and of the pampa as they needed more food. If I am correct, this wash would have been the first instance where local ditch builders confronted a topographic feature that required complex technical solutions; and the solution they devised and built would have prepared them for more difficult challenges the ditches to the pampa and to the Qolla Wayku would present.

These ditches, though, do not trace only the technological, demographic, and agricultural history of the landscape in which they were built, they also record the histories of conquest, colonization, exploitation, resistance, and political struggle. The present class relations in this small zone, and the ways they structure access to landscape resources there, epitomize the ways that the Spanish conquest and the dominion of the *hacendados* distorted indigenous relations with, access to, knowledge of, and stewardship over, the environment.

Mallma Alta is owned by two brothers, and constitutes the largest single holding that remains from the legacy of Cirilo Paz. Paz, economically and politically powerful, owned an hacienda which, at the beginning of the last century, included the uplands above the Qewar's source, all of Wataqasa, Mallma Alta, and Mallma Baja. That is to say, he controlled all of the land on the left, or north, side of the Qewar from above its source all the way down to the Ccorca. This land was divided among Paz's descendants, in lines from a son and a daughter, so it was not affected by the Agrarian Reform, even though it remains in relatively large private holdings.

Mallma Baja contains the large, flatter area below and east of Chuspi Grande. It runs from Ñekes, the ridge which projects from the high end of Chuspi Grande's east face, down to the Qolla Wayku, and along it to Paqpiri, where the Qewar joins it. In 1984, it was still owned by three different heirs of Cirilo Paz's daughter. They used their private property rights and their favorable location to use far more than their share of the waters in the ditch. The auxiliary ditches to these properties start high on its east face, soon after the ditch curves, out of the Qewar valley; so relative isolation and private property enhance its owners privileged access to waters. In addition, the soil here is sandier than the clay soils further east and south, where most of the smaller owners have land, so it can be irrigated more intensely with less danger of erosion.

The combination of sandy soils, privileged access, and extensive holdings allows the few large landowners north of the Qolla Wayku to use much more water than other farmers, and they abuse this privilege. I saw some of their fields irrigated over and over long before the small holders further down the ditch had irrigated even once. I also saw these fields being irrigated on days when users lower down had told me the water had been assigned to them that day. The relative proximity of Paz's heirs to the source of the Qewar facilitated their traveling to the ditch source to open it; it was predominantly these large holders who used the ditch before it was cleaned, that is, before the breaks had been repaired and the undermining of the walls by ditch water stopped.

Much of Mallma Alta has reverted to grasslands, and most of the cultivation done there is based on rental or sharecropping. Because it has not been ajudicated either to the Tihuicte cooperative or to the community of Maska, and thus is not formally under any community jurisdiction, its privileged position as the first major irrigable expanse of land under the ditch allows its owners to use a great deal of water. They let the water flow across pastures and down to the Qolla Wayku; their neighbors, and especially members of the community of small holding peasants further down the ditch, resent what they see both as a wasteful use of water and as the unjustifiable legacy of an anachronistic form of class power that deprives them of a much needed resource.

Their extravagant use of water appears ostentatious and abusive. Halfway across Wataqasa's east face, a major outlet releases water over a long waterfall to divide into two separate streams which flow across the bottom part of the grasslands above the Qewar. When they leave this outlet open, as they frequently do, it is possible to see the silver band of its voluminous flow from the pueblo, almost 2 km. away.

The inhabitants of Wiñay Poco maintain the entire 3,850 meters of ditch between the Qewar and the gullies above into which the ditch finally flows, but Mallma Alta's privileged access to and indiscriminate use of the water allows its owners to restrict Wiñay Poco's use of its water to small and unpredictable amounts for irrigation. A long history of violence and abuse by Mallma's owners, and the absence of authorities to control water use, discourages Wiñay Poco's residents from taking any action to assure their access to this vital resource, so their lands, poor and broken from years of overuse, continue to produce even less than they could with more water.

The social and the ecological history that Mallma Alta's and Mallma Baja's ditches etch into this landscape suggest some of the historical, as well as the geological and hydrological, dynamics that led to the collapse and abandonment of the other ditches. As we saw, the terrain here, though steep and, in one place, broken, was not as difficult either to build or to maintain ditches on as the higher routes over and around Chuspi Grande. The construction and layout of the ditch across Mallma Alta to Wiñay Poco were well adapted to equitable distribution of waters among a population motivated to broad participation in the sacredly mandated and regulated maintenance and repair of the ditches. The exogenous imposition of hacienda control over land and labor extinguished both the equitable distribution of water and the ritually legitimated community labor of maintaining the ditch or husbanding the resources it made available. Most of the Mallma Alta ditch was topographically simple enough that it did not require the close, technically competent, care that the upper ditch over Pukaqocha did, so it has endured to the present even while bearing testimony to the distortions and destructions that followed the colonially imposed abrogation of the indigenous system.

The Ditch to the Tihuicte Pampa [D]

The other ditch still in use starts lower, in the Qewar's main valley, about half a kilometer below its fall through the trees, at 3,579 masl. This ditch was almost surely built during the regime, and under the direction, of the *hacendados*. We can therefore examine its trajectory, its construction, and the history of its management to analyse how that system of dominion affected the relations between society and nature, as well as the technologies that mediate those relations. We will see particularly the ways that the hacienda's fragmentation of community and imposition of exogenous power simultaneously a) abrogated the authority and capacity of local leaders who had earlier mobilized the community through their knowledge of the environment and their technological imagination and b) undermined the community cohesion that had enabled the earlier, far more complex, careful collective labor of ditch maintenance.

Like the higher of the two abandoned ditches, this one serves to take water from the Qewar across the double barriers of Chuspi Grande and the Qolla Wayku gorge, but it follows a technically less challenging route, at a far lower elevation around Chuspi Grande's east end. Unlike the higher, earlier ditch, this one flows too low to irrigate any fields in the Qolla Wayku valley or any but the lowest slopes of San Miguel. Nor does it reach Llaspay. Its much easier route to the pampa limits its waters to the lowest agricultural zones around the pueblo.

Unlike the ditch above Mallma Alta, it flows through cultivated areas rather than above them, so maintaining minimal drop was an important means of maximizing irrigable areas. This consideration, combined with the topographical obstacles it was designed to overcome, made both the conceptualization and the construction of this ditch very elaborate. At the same time, the choice of easier, and sub-optimal, routes demands more knowledge of who organized the construction of the ditch and how. The most commonly told story about the origins of this ditch, affirmed by the more knowledgeable informants, was that Daniel Galdos, an *hacendado* and perennial mayor, ordered and coordinated its construction at the turn of the last century. As we did with the earlier built ditches, we will examine how the trajectory of this one fits into, and changed, the landscape across which it flows in order to learn more of the history of its construction. That this ditch was built relatively recently, though, is not open to doubt. I can assert with confidence that there would have been no reason to build this ditch before the higher one was abandoned, because the high ditch irrigated all the lands this lower one ever has, and then a great deal more.

After leaving the Qewar, the ditch flows east along Chuspi Grande's steep, sharply eroded, north slope. The contour the ditch follows averages about 12 meters of vertical drop over a mere 10 meters of horizontal run. The lower of the now abandoned ditches, which we followed around the top of Chuspi Grande earlier in this chapter, broke in multiple places above this slope, and the remains of that ditch continue to concentrate the flow of rainwater through those breaks, and then down steadily deepening gullies. These active gullies made the construction of the lower, more recent ditch more difficult, and continue to compromise its integrity. Its builders had to construct rock and earth retaining walls up to 7 meters high and strong enough to withstand powerfully erosive washes of rainwater on this highly eroded, distinctly unstable, ground. At some points, the builders cut the ditch across and into large rocks that reach as high as 3 meters above the ditch floor. At others, they built retaining walls onto the steeply sloping ground at the base of and against the rock.

The gullies presented a major obstacle in building the ditch, and continue to create major problems for its maintenance. It lies near the bottom of Chuspi Grande's steep high slope, so receives the full brunt, largest volume, and greatest speed of the water that runs off cultivated land on the top of Chuspi Grande into the breaks in the old ditch near the top, and is then channeled from this wider area into a few deep gullies. During the wet season, when torrential rains are frequent and prolonged, the ditch walls across these gullies must bear the force of fast, heavy streams of water which flow uninterrupted until they hit the ditch. There is evidence of repeated breakthroughs in the walls at these points, and the heavy streams of water pouring over the walls into the gullies below have undermined the retaining walls. At two points in this section, it is possible to look straight down from the ditch and not see the retaining wall below. At one of these points, the washout under the ditch is 4 meters across.

A seriously inadequate system of ditch maintenance aggravates the problem. The ditch is cleaned and repaired in a day of *faena* (collective task) labor by members of the Maska community and of the Cooperative, as both groups use it. The legal complexities of the Agrarian Reform distinguished different persons' rights in land in terms of their earlier relations to particular *hacendados*, and then subjected the resolution of claims on each hacienda through costly and complex legal processes (Seligmann, 1987). The Reform thus created different groups with different rights in land and different political and economic regimes of collective and of individual production and labor. In Huanoquite, the resulting divisions have been particularly problematic between the Tihuicte cooperative and the Maska community.

Community and coop are at odds on a series of issues, and each has its separate set of authorities. This rivalry is complicated by ambiguities about whether the Consejo, the governing body for the entire district of Huanoquite, or the authorities of Maska, which comprises only a part of the pueblo of Huanoquite, are in charge of the ditch. Finally, there is great inequality in access to ditch waters. A very partial agrarian reform left large pockets of privately held land, much of it of excellent quality, under the ditch (Seligmann, 1987). As we have seen, some haciendas divided between multiple heirs over several generations were not appropriated or adjudicated to the community because their individual parcels, though considerably larger than median holding sizes, were not large enough to merit the huge cost to the community of the ultimately uncertain legal processes of reclaiming them. Furthermore, the land that was adjudicated to the community and to the cooperative were not equally distributed among their members, and some members of the community have inherited or acquired private parcels as well. Differences in political status and authority within the coop seems to have influenced its different members' access to land. Despite the community's ideology of equal rights and obligations, its lands are apportioned according to individual capacity to work the land. In the community, therefore, the more prosperous, who can hire more labor and who have more animals for plowing and transport, receive more land. These members frequently have their own private holdings under the ditch, while very little community land is presently irrigated.

The cooperative divides its lands more equitably, but even here, more powerful members often are allotted greater portions of the collectivity's lands. In addition, most of the cooperative's irrigated land is reserved for collective, rather than individual, cultivation. Finally, and perhaps most important, the cooperative is governed by rules established by the national state that appropriated the haciendas in the first place and that imposed this exogenous form of authority, mobilization, and distribution of resources as a condition of the ex-*feudatarios* gaining rights to this land. The office-holders and authorities of the coop, then, no matter how democratically elected, are inevitably seen, especially in moments of tension or conflict, as personifications of an alien state, and, at one step removed, of the alien *hacendado*. This perception became particularly acute as organizational and managerial problems in Tihuicte led most members to feel that they are not sufficiently rewarded for the labor they devoted to the cooperative's crops.

These diverse forms of inequality, and sources of resentment, preserved, transformed, and exacerbated the hacienda regime's fragmentation of the community cohesion, solidarity, and trust that had nurtured a) the sensitivity to the environment, b) the technological and topographic knowledge and imagination, and c) the unanimous and sustained mobilization of the multiple communities required to build and to maintain the original high ditches. This fragmentation sharpens the contradiction between rhetoric and practice. Despite all of the inequalities in individual use of and benefit from irrigation water, the ditches and the water they carry are defined as a public property and a collective good, and all members of the community and the cooperative are expected to collaborate in maintaining the ditch. Divisions within and between the various collectivities mean that the authorities in each have trouble coordinating their efforts and dividing the labor involved in what has to be performed as a collaborative effort; they also have trouble organizing and mobilizing their respective members to get out to do the work. In 1984, the ditch cleaning *faena* took place over a month later than I had been told it would. In the meantime, people had been using the ditch, and water had been spilling over the breaks, deepening the gullies that undermine the ditch walls.

When the *faena* did take place, 125 men turned out, a large number, but still only half of the men who should have been there. Some of these were young boys substituting for their fathers. The work went fast, but there was little overall supervision. As a result, the breaks in the walls were built back up enough to prevent water loss during the irrigation season (August-November, though in this year August was already past), when it rains little or not at all. No structural work at all was done on the retaining walls, nor were the breaks repaired back to the height or width of the rest of the ditch. Thus, though the walls along most of this section range up to 145 cm. above the ditch floor, and their width up the 225 cm., the repaired breaks were only about 25 cm. above the ditch floor and about 35 cm. wide. These would wash out quickly and easily in the following rainy season.

When I talked to various people, including past and present authorities, about this problem, they agreed that the ditch was deteriorating here, but said either that people would no longer comply fully with their *faena* obligations or that concrete reinforcement was needed and should be provided by the state. Answers to my questions about how the work had gone before the Agrarian Reform were more varied, but a frequent response was that the *hacendados* often failed to comply fully with the requirement to supply a full complement of workers for the *faenas,* and that they often failed to coordinate their different work forces adequately. Some specifically mentioned that the division of both Galdos's and Paz's haciendas between multiple heirs made coordination and collaboration on the ditches, and effective regulation of their use, much more difficult to achieve long before the Reform. They also related this problem to the accelerating deterioration of the ditches

The stories about particular *hacendados*, the physical evidence that extensive erosion and undermining preceded the collapse of the high ditch, and the myths and legends Huanoquiteños recount about the ways their ancestors interacted with land and water, all strongly suggest that the hacienda regime was not particularly effective in mobilizing either the knowledge or the labor required for adequate ditch maintenance. Certainly, the multiple signs of breaks, and of careless repairs, associated with clear evidence of serious erosion under the ditch, suggest a longer period of neglect than the short period that had elapsed since the reform when I first saw the ditches in 1984; so the organizational and distributional problems after the reform can be seen as new manifestations of earlier, and persistent, contradictions. These evidences corroborate my argument in the last section, that it is difficult to imagine a coerced, resistant, or resentful labor force working with the attention, concern, and effort that maintaining this complex ditches would require.

Though far less spectacular than that of the old high ditch, the original construction of the ditch was as impressive as its present maintenance is deficient. Despite the steep slope of Chuspi Grande itself, the large boulders and rock faces which had to be moved, cut through, or traversed, and the deep gullies that had to be crossed, the ditch builders managed to lose only 7 meters of altitude in the ditch's first 780 meters, with 1.4 meters of this drop occurring in the first 80 meters of distance, where the fall had to be sharper to augment the amount of water captured from the river. Though the retaining walls are not carefully maintained, they lasted beyond living memory, resisting time and the force of water, despite their steep and unstable bases. One possible interpretation is that Galdos was a powerful and effective organizer, but that other authorities since were far less competent or charismatic, and were thus less successful at overcoming the multiple sources of fragmentation and resistance that the hacienda regime had created.

At 780 meters from its source, the ditch turns around the end of Chuspi Grande, already 36.6 meters above the Qewar as this river reaches the pass out of its valley between Chuspi Grande and Wataqasa. There is an outlet here, and the auxiliary ditch down the slope coincides with a foothpath, enhancing erosion. There is a major washout below the ditch, over 30 meters wide and very unstable. There is a retaining wall under the ditch, with a nearly vertical drop of 7 meters just below it. Again, the ditch is very shallow here; its lower wall does not exceed 25 cm. above its floor along the entire 60 meters around the curve.

From this curve, the ditch runs almost due south along Chuspi Grande's somewhat gentler (20-30 degree) east face, with only slight curves to accommodate washes and other irregularities. Almost all the land on this slope is cultivated, and there are numerous outlets to small canals, many of which follow paths and gullies, some of them continuing over half a kilometer away from and 200 meters below the ditch to the fields above the Qolla Wayku's left (northwest) bank.

The fields on the upper end of Chuspi Grande's east face were adjudicated to the community of Maska after the Agrarian Reform, but these lands are effectively controlled by 6 community members who had farmed them in return for labor obligations to the former *hacendado*, so that even though they officially belong to the community, they are used as if they were still private property. The rest of the land below the ditch on Miska Pata is still owned in two large private holdings at the bottom end of this face, that is, above the end of the Qolla Wayku's gorge.

Despite the intense cultivation, Chuspi Grande's east face is not as gullied as its north slope. Irrigation has had a distinct effect, however. Looking at the east face from the pampa, one can see a clear change between the smoother slopes above the ditch and more broken slope below it. Also, inspection of individual plots shows clear signs of erosion after plowing and irrigating. The paths and gullies that the ditches follow have all been significantly deepened by the water's flow. Interestingly, though the course of the abandoned ditch at the top of Chuspi Grande passed high above the one now in use, the upper slopes between the two ditches have not suffered much erosion because, through this section, the upper ditch remained relatively unbroken. The severe erosion below the older ditch on Chuspi Grande's north face however, is clearly associated with ditch breaks on this much steeper slope. The upper slope does become rocky and barren at the south end of its face, toward the Qolla Wayku gorge, but the large cirque-like hollow there appears to be the result of a major landslide, perhaps provoked by earthquakes, rather than of the gradual erosion characteristic of cultivation and irrigation.

During their first 700 meters south across this face, the builders continued to drop the ditch slowly, losing 10.2 meters of altitude. Construction was easier here than on the north face; at many points the lower wall of the ditch bed was simply built up on the slope. Outlets to small auxiliary ditches are usually framed with rock, and rock is used to close them. Nowadays, during the irrigation season, there are numerous leaks and filtrations from these outlets, the result of hasty or careless closing by individual users and of the absence of close supervision by their neighbors or by authorities; but these leaks are slow, and do not cause the damage which the major breaks above the Qewar do.

When the builders turned east-southeast into the broken cirque-like hollow near the Qolla Wayku gorge, however, they chose to drop more rapidly, staying near the bottom of this rocky, steep, unstable formation. The terrain is far more difficult here, with rises above and drops below the ditch of up to 16 meters vertical to 10 horizontal meters. The ditch goes over several waterfalls, and drops 12 meters in its next 100 meters of flow and 6 more in the following hundred, that is, more than it dropped in the previous 1.5 km. Here as well, it crosses over retaining walls, and the steep outlets to the fields below have undermined these walls.

The ditch reaches the end of this hollowed out area in about 250 meters. Here, the builders turned sharply south-southwest into the steep narrow gorge of the Qolla Wayku, running the ditch up the gorge, against the flow of the river below, across a sheer rock face, which in sections is vertical for 7 meters above and 13 meters below the ditch and whose overall slope exceeds 60 feet. Here the ditch is held up by the most impressive of its retaining walls, built on bases so steep and unstable that I, carrying no rocks and certainly not attempting to stack them permanently on top of each other, slipped and fell several times just trying to get close enough to examine their construction. I can only guess that these walls go down to foundations well below the steep sandy ground I was scrambling across.

The ditch flows south-southwest across these high retaining walls for 200 meters, meeting and crossing the Qolla Wayku river at a point where the base of the vertical rock face intersects with the narrow (2 meters) floor of the river. The river drops sharply just below this point, and the ditch builders constructed a rock wall across the river bed here, just where it starts to drop, laid long reeds across the rocks, and then built the ditch bed itself out of *champa* (sod). The reeds and sod are reinforced and partially replaced in the yearly *faena*, the workers bringing the large blocks of *champa* from the last field before the ditch starts into the broken hollow, about half a km. away. A short ditch diverts much of the Qolla Wayku's water into the ditch; the rest filters through the rocks under the reed and sod trestle and continues down the gorge. The ditch here, 1,932 meters from its source to the Qewar valley, has dropped 41.5 meters, to 3,537 masl. (24.3 meters of this drop occur in the last 500 meters of flow).

I explored up and down from the ditch along the Qolla Wayku's bed, and could not find a more favorable point for crossing than where the ditch is located. Above the trestle, the right wall of the gorge, along which the ditch has to travel back out again, is composed of long 70 degree slopes of small rock mixed with sand, much of which has already slid and all of which continues unstable. The gorge and the river bed through it both become even steeper below the trestle. Situating the trestle any further downstream would have reduced irrigable land further down the ditch.

The ditch doubles back out of the Qolla Wayku gorge across slopes almost as steep as the ones on the other side, along which it entered. The fall below the ditch reaches 20 vertical to 10 horizontal meters. Again, steep retaining walls support the ditch; these walls, unlike those on the other side of the gorge, have been severely undermined by erosion at several points.

Sixty meters beyond the trestle, the ditch curves east out of the gorge across the steep, rocky, deeply eroded slope (35-40 degrees) where the north end of San Miguel falls to the Qolla Wayku. There is no cultivation above the ditch here and only a few small fields on narrow shelves about 40 meters below it. The outlets to these fields go down the sharp gullies and thus erode them more deeply. The ditch here continues to require retaining walls, especially across the gullies but also along the very unstable rock and dirt base of this steep hillside, which it follows for another 750 meters.

Despite this difficult trajectory, however, the ditch drops only 19.5 meters, to 3,518 masl., in the first 950 meters of flow past the trestle. Beyond this point the slopes above the ditch become a little gentler, and there are a few small fields cultivated here. The drop below is quite steep, but over half of this slope, which reaches 30 degrees in parts, is cultivated as well.

The transition between the steep, unstable, uncultivable slopes near the Qolla Wayku gorge and the still steep, but more stable, partially shelved slope the ditch now starts to cross, about a kilometer beyond the gorge, is marked by Lambranwayku, a stream that has cut a deep, dramatic, boulder-lined bed into the flank of the hill. The ditch crosses Lambranwayku, and then a secondary gully into it, on a curving wall that runs for over 100 meters. After Lambranwayku, the ditch passes through the extensively terraced, intensely cultivated fields close to Huanoquite. Water and land are both directly affected by the denser populations and social activities. Multiple small feeder canals run from the ditch to the house-gardens and small fields associated with different houses in the village. Terraces, fences, and paths create myriad intersecting lines. There are still visible traces of a cobbled Incaic road going up Lambranwayku to Wankankalle. We will see as well as we follow the ditch across these more densely settled, more intensely farmed, spaces how different systems of tenure and of power, imposed during different historical periods, continue to mark both the landscape and the social relations within and between the collectivities that transform and exploit that landscape.

San Miguel's slope extends further out here; this increases the distance between its crest and its foot and allows the wide shelves on its slopes above the pueblo. The ditch runs north for 120 meters through this section. There are several outlets in this stretch, most of which carry water all the way down to the triangular terraced shelves, intensely cultivated, formed by the confluence of the Watanay with the Qolla Wayku, 200 meters below and 500 meters north of the ditch, and just west of and 50 meters below, the pueblo itself.

The ditches from these outlets also follow gullies down the steep upper slopes. In some places, the force of the water has cut over 50 cm. into the large sedimentary rocks after water eroded the topsoil off of them. As the slope diminishes near the fields, a series of small lateral ditches running along footpaths and under rustic terraces (see Fonseca, 1984) distribute the water to the various fields. These terraces, mostly of sod, some reinforced with maguey, have further leveled what appear to be naturally occurring shelves or breaks in the steep slopes. Human intervention here has combined water and level areas on warm, sheltered, fertile land close to the pueblo to create an extremely favorable, though small, zone for the cultivation of corn and early potatoes. Most of the plots here belong to ex-*hacendado* families.

The main ditch drops sharply in this section, over 4 meters in 120, to the place called Chinku, a hill overlooking the pueblo, 3 km. from the ditch's source. Outlets here carry water down a low ridge that projects from the hillside over the west end of the pueblo. Forty-five meters beyond this point, the ditch flows around the head of a long, wide wash, also called Chinku. Here the builders constructed a retaining wall of *champa* which reaches 3 meters of height and continues for over 100 meters. A major outlet carries water down this wash, which becomes a sharp gully at its bottom, into the west end of the pueblo and to the fields below it above the left bank of the Watanay.

After crossing the Chinku wash, the ditch continues southeast for 830 meters, in two wide, shallow northward curves between two other washes, across the gentle middle slopes of San Miguel. The ditch divides these slopes into two distinct shelves; the upper one long and flat (5-8 degrees), ranges up to 200 meters in width, but is diminished to half that width by a second wide, shallow wash which flows down toward the middle of the pueblo 400 meters past Chinku. The lower shelf is steeper (up to 30 degrees). It drops from just below the ditch to the pueblo. The ditch, which flows above sod retaining walls that reach 4 meters in height around the second wash, but are mostly about half that height, appears to form a terrace between the upper shelf and the lower slopes. It is conceivable that the ditch was built along an existing terrace, or that years of cultivation above the ditch loosened soil which filled in above the ditch. The debris in the ditch is thrown uphill during the annual cleaning, thus heightening the levelling effect.

The ditch floor is often as much as 2 meters below the bottom of this shelf, and even though the shelf itself is quite flat, the rise just above the ditch is usually as much as 4 to 5 vertical over 10 horizontal meters, suggesting that the lands above the ditch may have filled in somewhat over time. The slope below the ditch, however, is usually between 5 to 7 meters over 10, and, for one 80 meter stretch, increases to between 10 to 12 meters over 10. This, together with the much greater overall slope between the ditch, would indicate that even if some of the levelling above the ditch has resulted from cultivation since the ditch was built, there was already a clear difference between the slopes above and below it. In other words, the builders constructed their ditch at the lower edge of a relatively flatter slope, before it began to drop more sharply.

The entire face of San Miguel above the pueblo was once part of the hacienda of Daniel Galdos, who also served several terms as mayor at the turn of the century and is said by some people in Huanoquite to have organized the construction of the ditch we are now describing as well as a smaller one, higher up on this same slope that waters the flat upper shelf. This land has now become fragmented among several generations of Galdos' heirs and people they sold to. The shelf above the ditch is part of Llulluchayoq, a large holding acquired from Galdos' granddaughter, and the slopes below it have been divided among heirs of subsequent generations. Though these plots are numerous and small, a significant number of them are owned by families who continue to be wealthier and more powerful than most Huanoquiteños. The upper shelf and the slope below are both intensively cultivated and highly productive.

In addition to the numerous small outlets to these fields just below the ditch, several larger canals follow both the wash and smaller gullies down to the pueblo. The water from these canals flows through the pueblo in small ditches, down streets, and through patios to water house gardens and the small fields on the left bank of the Watanay. Many of these ditches go through small holes built into the adobe walls around house compounds, dropping down to patios or streets lower down on the gradual slope at the foot of San Miguel. Various of the walls which hold patios and house-gardens level also have drainage holes built into them, 1 or 2 meters below the level of the ground inside. These outlets in these walls, which are themselves very old, suggest that irrigation here may be as old as the pueblo itself. Water is released in these ditches both for cultivation and for making adobes. Sometimes the ditches start to flow in the evening, when people stop irrigating their larger fields above the pueblo and the extra volume of water overflows the closed outlets.

The ditch falls slowly across San Miguel's slopes above the pueblo, dropping less than 12 meters in the 920 beyond Chinku. At this point, the ditch reaches Huanakauri, the sacred site that we encountered earlier in this chapter while we followed the abandoned ditch that crossed this same slope higher up. At the lower level we are following now, just beyond Huanakauri, the builders had to carry the ditch across a deep gully that runs northeast from higher up San Miguel.

This gully marks the abrupt end of the flat, fertile shelf above the ditch. Beyond it, to the southeast, is Tanqarpata, the rocky, arid, severely eroded ridge of San Miguel which slopes down over the southeast end of the pueblo. The soil changes here, from a clay to a sand base. Instead of the deep rich topsoil mixed with sedimentary and volcanic rock, some of these so altered that they crumble when one twists them, Tanqarpata is covered with large bare crags and boulders, all volcanic, probably not as old as the rock base of the shelves.

Just before the ditch reaches Huanakauri, a major auxiliary channel crosses a wash that runs between Huanakauri and the pueblo. Sod retaining walls bear this channel across to a ridge. The channels flow along the ridge down to the pueblo, thus maintaining greater elevation than is possible in the majority of auxiliary channels, which follow more steeply descending gullies. This added elevation permits the irrigation of extensive areas at the south end of the pueblo.

Huanoquiteños are aware of the difference in soil types and stability on the opposite sides of these gullies; they associate Huanakauri itself with protection of the pueblo, and the gullies as related to the powerful and dangerous San Miguel, at the same time that they associate the entire sector beyond Huanakauri with the Incas' power over rocks. The slopes above the ditch here are too steep and unstable to cultivate, and the cultivable areas further along are now too high to reach, given the drop in elevation that the ditch builders conceded in their passage around Qolla Wayku and then again at Chinku.

The builders changed their strategy for the ditch trajectory as abruptly as the topography changes here. From Chinqu to Huanakauri, every possible effort was made to keep the ditch high and thus maximize irrigable space. Below Huanakauri, the ditch follows its easiest trajectory. Here, almost 4 kms. from its source on the Qewar, the ditch falls 4 meters over a waterfall into the gulley, down the gulley for a few meters, and then along its right wall, held up again by sod retaining walls built into the steep slopes (up to 10/10 meters above and 8/10 below, with parts of this in vertical drops of up to 5 meters). About 200 meters past Huanakauri, the ditch turns sharply southwest, along the walls of an even steeper gully down Tanqarpata. This gully continues down to the pueblo to become the Watanay River. The ditch then turns back northeast, clinging to the side of the slope south of the Watanay, losing altitude rapidly as it goes over a series of waterfalls, built into its retaining walls. The ditch could have gone higher across the ridge, but the land above it is poor and unstable, and the extra difficulties of construction would not have increased significantly the irrigable area below it. Even so, the retaining walls in this section of the ditch, ingeniously built into Tanqarpata's unstable ground, are badly undermined. The steep drops below the ditch, and the loose soils of the slope it crosses here, together with the rushing of water over waterfalls within the ditch bed itself, all make its walls highly susceptible to erosion.

The repairs necessary to maintain the ditch across this crucial and difficult stretch, however, are even more difficult to organize than in the upper sections of the ditch. The community and the cooperative are supposed to collaborate in the one day *faena* as far as Chinku, just before the ditch starts across the bottom of Llulluchayoq's flat shelf. The community members alone are then expected to do the repairs across the shelf, where the ditch is relatively stable and the work is easy, and then on down to the base of Tanqarpata, where the coop lands begin. Most of the lands cultivated by community members, however, lie under the ditch above Huanakauri, so they are not very interested in the more difficult, but far more important, job of ditch repair in the steep gullies below Huanakauri.

Seven hundred and twenty meters beyond and 56 meters below the waterfall at Huanakauri, the ditch reaches the bottom of Tanqarpata, just above the buildings of Tihuicte, the hacienda which owned the pampa and is now a cooperative. At this point, 3,444 masl., a major ditch branches off to the left, traveling on a raised sod bed between Tihuicte's compound and a marshy, reed filled lake, Totora, which lies just southeast of it. This ditch drops rapidly to the north, down 30 meters in 342 of flow, until it meets the truck road to Cuzco on a rise that connects the pampa to the south end of San Miguel. This ditch then swings back northwest, traveling along the southwest and part of the west flank of the pampa falling from 3,414 to 3,406 masl., for another 1.5 km. It waters the small fields below it, above the north bank of the Watanay, as it flows back past, and above, the pueblo.

The main ditch continues east-southeast at the base of the Tanqarpata for another 300 meters. A smaller ditch continues east-southeast, following rapidly as Tanqarpata's base drops down to less than 3,200 feet toward the Ccorca. This small ditch served haciendas and small holdings further down until the agrarian reform, but it is now difficult for the small farmers in the lower areas to reach agreements with the Huanoquiteños, and the ditches now seldom carry water. The main ditch, however, turns northeast over a similar raised sod bed to cross the other end (southeast) of the lake. It then flows along a rise, 4 meters higher than the one that the ditch at the other end of the lake crosses. This carries it across to the pampa at 3,417 masl. It then follows the southeast rim of the pampa for 500 meters, dropping only 3.5 meters in this distance. Numerous auxiliary ditches flow off to the right here, to water the series of wide flat shelves which continue halfway down the pampa's southeast and east flanks as they drop to the Ccorca 450 meters below. The first of these shelves is Urubamba, a wide flat tongue of land which projects .75 km. southeast from just under the ditch, narrowing from 400 to 100 meters wide, and dropping only about 32 meters, from 3,410 masl. to 3,375 masl. A rock promontory with the remains of a stone watch tower or platform and a deep, straight channel, 20 cm. wide cut into one of the base rocks, rises at the southeast end of Urubamba. Like many other ancestral constructions in Huanoquite, it appears to stand guard over a particularly favorable agricultural area, in this case especially suited for corn, and also to provide a view of trade routes. The channel into the rock, though, is reminiscent of the grooves in the stones at Inkaraqay. They suggest ritual rather than defensive purposes—specifically, rituals celebrating, thanking, and imploring water. Ritually, cosmologically, and practically, water enables agriculture, the trade that agriculture makes possible, and the extended community that agricultural exchange across multiple ecosystems engenders.

Urubamba's gentle slopes are broken by beautifully contoured sod terraces which, in addition to levelling land and marking the boundaries between the small, privately held plots here, provide walls for the network of small auxiliary ditches which carry water to each of the plots. The terraces form elaborate curves, as they are fitted both to the southeast slope away from the pampa and to the northeast slope of the depression in Urubamba, which drains into a bordering wide wash. The ditches follow the terraced contours of both slopes. Urubamba is so flat that many of the terraces have drainage ditches at their lower ends as well as irrigation inlets above, so the ditches serve a double function, bringing water in and getting it out. The clay soils here make this double function especially important, as excess water takes many days to dry out and threatens to rot roots.

At the rim of Urubamba, the hill drops off at a 40 degree angle, but even this slope is broken by tiny shelves, not more than 10 meters wide, with traces of old terraces. At the bottom of this steep drop is Haqoyke, a 300 meter wide band with a series of small, relatively flat fields between 3,000 and 3,100 masl. Ditches along the rim descend slowly to complement the spring that waters these fields, whose lower altitudes and protective hillsides provide the most favorable agricultural conditions in Huanoquite for crops like corn that need warmer temperatures and longer growing seasons.

Beyond Urubamba, the ditch turns due north for 130 meters. It then turns sharply west for about 60 meters, running above a steep drop into a wide flat wash called Pukaqocha, where the cooperative grows corn and wheat. The ditch curves back north-northeast on sod retaining walls across the gully at the head of

Pukaqocha, at 3,370 masl. This same gully carries irrigation water down to Pukaqocha. The ditch continues under the nearly vertical walls of the low promontory which rises above the wash.

Here, for the first time in the ditch's course across the pampa, one can see the limestone and chalk base under the thin layer of soil, and one can see why the pampa, though quite flat, shows such extensive erosion, and why ditch maintenance and repair is so critical despite the apparent ease of irrigation. The ditch itself has been cut into the limestone base of this promontory, and the ditch water has dissolved the chalk veins in the limestone, creating deep (3 meters), wide (4 meters) sinkholes in the ditch's floor. The water then runs out below the retaining walls, weakening them and deepening the gully.

The cooperative's members have laid cement tubes on logs placed across one 16-meter series of sinkholes, and have bent tin sheets to form a bed across another. These solutions are costly, and have to be redone or reinforced almost every year. The ditch drops over 4 meters in 200 as it negotiates its way across this promontory, as opposed to a drop of only 3.6 meters in its first 600 meters across the pampa, above Urubamba.

Just north of and above Pukaqocha, a long wide ridge called Chincheromayuq (3,370-3,400 masl.) extends almost half a km. east from under the ditch. A major auxiliary canal carries water to large fields where the cooperative members collectively cultivate corn and barley. At the north edge of Chincheromayuq, at about 3,375 masl., are two smaller areas, Rakaypunku (in Quechua door of the ruined house) and Marca Wasi (Wasi means house), where informants told me there were previously Incaic ruins. Directly below Rakaypunku is the extensively terraced slope of Saqsi (3,100-3,200 masl.), a major corn growing area held in private plots. Saqsi is bounded to the north by a rocky promontory called Bandon Wajana, and to the south by a mound topped by two concentric stone walls, called Saqsi Muqu. These two hills are within calling distance of each other. Between them, they command a view of a broad section of the Ccorca valley and of the valleys which flow into it east of the pampa and down toward the Apurimac. These too could have been lookout towers over approaches to Huanoquite and over the cornfields on Saqsi's terraced slopes, or they could also have served ritual and celebratory functions. Urubamba and Saqsi both overlook areas that were linked to Huanoquite by trade, kinship, and community, rather than opposed to or competitive with it militarily.

In addition to all of these indications that the south and east flanks of the pampa have been intensely cultivated since pre-Incaic times, one can look up at the pampa rim from Saqsi and see the extent to which the present contours of the pampa have been formed by erosion into gentle swales between rock promontories, with much deeper gullies cutting down the floors of the wider swales. These erosion patterns suggest, but by themselves cannot confirm, long years of continuous cultivation, and perhaps of irrigation.

Auxiliary ditches reach Saqsi across Pukaqocha and Chincheromayuq, and then travel laterally along terraces. Irrigation is important here, but it is also increasingly difficult, as the water has deeply eroded the gullies that serve to carry the water downhill. Similar erosion occurs in the ditches that drop down the pampa's flanks to lower shelves and fields. A ditch which goes over the east end of Urubamba toward Pukaqocha, for example, has so eroded the underlying limestone and chalk base that the ditch-bed has collapsed, and water runs underground for 250 meters, making irrigation impossible in this stretch.

In addition to the problems of gullying, which lowers the ditch bed so much that it is hard to get water out of it to the fields, the owners of the fields at these low elevations must bring the water through a maze of small ditches and terraces. As they are literally at the end of the line, they are susceptible to many other users higher on the ditch taking the water before it reaches them. The owners of these fields thus have the added difficulty and labor of patrolling the long extensions of ditches above them, and may have to fight to keep their water, even if it is supposed to be their turn to use it. Many of the owners of fields further below Saqsi don't even bother to irrigate, but wait until the rains come before planting; an option that the longer growing season and the freedom from frost at these lower elevations allows them. Still, their dilemmas, and their giving up most hope of irrigating because of the high personal cost of assuring their turn and a fair share, epitomize the consequences of the fragmentation of community discipline and cohesion, and the dissolution by competing factions and interests into an ongoing tension between individual and communal rights and obligations. The precariety of individual rights to water generates, replicates, and exacerbates the tenuous claims that any of the contemporary collectivities have on the intense and unanimous community labor and care that the ancient ditches still bear physical witness to.

Continuing north, the ditch flows above Cruzpata, a wide flat area just down from Chincheromayuq, which also belongs to the cooperative but is primarily cultivated by individual members. It then bends west into a sharp gulley called Kuchillo Wayku, again flowing across sod retaining walls. Here, the present ditch drops rapidly to avoid a sheer limestone face on the other side of the gully, but there are traces of an older ditch that crossed the gully about 4 meters higher. Cultivation beyond this point has eradicated all traces of this old ditch, but it seems likely that it washed out as its waters dissolved the chalk in the limestone base of this steep hillside, and was replaced by the lower, easier course of the present ditch. The sharp drop here (6 meters in 200) sacrifices a significant amount of cultivable land which the old ditch would have watered further north.

The ditch flows north-northeast out of Kuchillo Wayku, finally dropping below 3,400 masl. Then, 7 km. from its source, it curves northwest for 100 meters into another gully, which widens out into a large flat field 6 meters below the ditch. The left hand side of this area, Zanjapata, is farmed collectively by the cooperative. Individual cooperative members cultivate smaller, somewhat steeper fields on the other side. These areas are all irrigated by auxiliary ditches which flow down the gully.

The ditch flows northeast for 400 meters past Zanjapata, along the slopes of Wilkuru, which are cultivated by individual cooperative members. The ditch has become quite small now, about 25 cm. wide and 20-30 cm. deep instead of the full meter of width and half meter of depth that it maintains for most of its course from the Qolla Wayku trestle through the first half kilometer across the pampa. The ditch continues to drop rapidly, 10 meters in 200 at the head of Zanjapata, and 6 in 400 along and around Wilkuru.

At the end of Wilkuru, the ditch turns west for another 400 meters, passing above Chiwarpampa, a long, narrow, broken shelf perched above the precipitous drop from the pampa's north slope down to the Ccorca. On this side, there are only a few very steep, small fields below the pampa's rim, and all of these are nestled into the sides of the gullies which plunge over the edge, breaking and opening somewhat the walls above the river. Slopes become even steeper toward the bottom, and cultivation there is impossible.

The ditch presently ends abruptly in a deep gully, though there are a few traces on the other side, so it appears that it contributed to the erosion which eventually broke it here. The presently useable ditch ends at 3,379 masl., 7 kms., 860 meters of flow and 200 meters below its source on the Qewar. It has flowed from east to south to east to north to west, with numerous curves interrupting its course in each of these directions, so that its end point is only about 3.4 kms—or less than half of the total distance it has flowed—east and slightly north of its beginning. It has crossed a landscape that incorporates the palimpsest of the radically different systems of property and power that have been imposed on Huanoquite since the 16th century, flowing through lands acquired and held through very different tenure relations. Some belong to the community of Maska and are granted to individual members for seasonal use, others are privately held in relatively large parcels by the heirs of *hacendados*, and others are broken into small private plots acquired commercially or in exchange for labor. Some of what the cooperative members acquired from the *hacendado* they were bound to they now farm collectively; other parts of the pampa they farm individually, but scattered within and below the pampa are small parcels of privately held land.

The ditch's trajectory across the limestone base and the varied tenure systems of the pampa, the physical work and the social relations of maintaining and repairing it there, and the uncertainty that taints any member's or land-owner's certainty of access all underscore, and simultaneously exacerbate, the complexity of maintaining a ditch that provides water for cultivation at altitudes ranging from over 3,570 masl., where the cold limits cultivation to potatoes, wheat, barley, and leguminous grains, down to less than 3,100 masl., where warmer, more protected fields provide favorable conditions for corn. By the time the ditch ends on the pampa, it has crossed the slopes of two major mountain formations, traversed a deep gorge and numerous sharp gullies, and traveled for long stretches of its course over retaining walls necessary to get it across steep and unstable soils. Its builders sustained minimal drop even across difficult obstacles in order to maximize irrigable area, but when the land beneath the ditch would have been very unstable, or excessively difficult, or when the land was not particularly good, as on Tanqarpata, the builders sometimes allowed the ditch to drop, sometimes over waterfalls they built into its course. They took advantage of the prevailing slopes of different hills, and took multiple other topographical features into account, but they also invested great ingenuity and huge amounts of labor in building walls and trestles and in cutting through rock, that is, in changing the physical base sufficiently to hold and direct the ditch waters.

The ditch carries up to 110 liters of water per second to an irrigable total of over 4 square kilometers. Even though much of this water is lost in filtrations, badly closed outlets, breaks, and negligent waste, the ditch allows corn, potatoes and other minor crops to be planted during the dry season, and thus to mature before the frosts come at the end of the wet season. It thus significantly changes the agricultural options available to the Huanoquiteños.

It also creates and accentuates inequalities between different individuals and groups. The community as a collectivity has little irrigable land, and what it does have is effectively controlled by 6 of its members. The cooperative has a great deal of irrigable land, although it lacks the high puna lands which the community members use to grow potatoes. Private holders control most of the irrigated land above the pueblo, some of these are also community members who receive disproportionate amounts of land because they are better able to organize the labor necessary to work it. To some extent, their ability to organize labor depends on the amount of corn they can grow, because the *chicha* made from corn is a major part of labor expenses. Irrigation is, of course, essential for growing corn in most of Huanoquite, and so irrigation provides resources critical to the maintenance of the power and status that translates into privileged access to water.

Contemporary Huanoquiteños work hard at irrigating, and they spend two days of heavy collective labor every year cleaning and repairing the ditch. The major, most difficult, and technically complex work on these ditches, though, was done years ago, and the assault on and degradation of the communal system that sustained the ecological and structural integrity of the irrigation system has been progressing for many centuries. The ditch to Mallma Baja may have been built by indigenous authorities in a highly egalitarian system, but it had been appropriated and abused by *hacendados* long before the Agrarian Reform failed to reverse the inequalities of the hacienda regime. The same regime catalyzed the neglect and abandonment of the higher ditches that had distributed water much more equally over a far greater area, and then had commanded the construction of the lower ditch we just followed across a far smaller irrigated area. Now, ongoing fragmentation of the community threatens even the integrity of the system that *hacendados* ordered built. Key parts of the most difficult and elaborate sections of this ditch are deteriorating, and irrigation water is hastening the erosion of various cultivated hillsides. Unless the Huanoquiteños find some way of overcoming the economic and political obstacles to the coordination of individual interests with public needs, this deterioration will continue, and will eventually have disastrous consequences for their agriculture, repeating the earlier collapse of the ancestral ditch.

After describing briefly several minor ditches, I will consider the extent to which, and why, the ditches that the Huanoquiteños received from their forefathers constitute an environmental modification that they still know how to use but may not be able to maintain or to replace. I will then ask what we can learn from them directly, from their myths and rituals indirectly, and from their ancestors through the medium of the landscape they modified, about how some individuals' communication with earth and water transform society's relations with nature, and how belief systems, social organization, and physical practice in turn are shaped by those relations.

Chapter Four

Power, Property, and Technology

In this book so far, we followed all of the major ditches—still in use as well as abandoned—from their sources to their end-points. I showed you how their builders fitted each one into the natural interactions of rivers and land-forms. In this chapter, I continue our consideration of how different regimes of property and power might have affected Huanoquiteños' means of imagining, building, and maintaining these intensely formative components of the landscape they depended on for their own survival and procreation. I discuss the ways I examined other modifications of landscape for additional sources of information I used to understand how earlier Huanoquiteños might have imagined and organized ways to construct irrigation systems. I also consider what more these other landscape modifications reveal about how political systems dominant when they were constructed would have affected environmental awareness and technological imagination. I then recount the various stories and legends about the ditches. Finally, I discuss their meanings, and how those meanings changed with historical changes in the regimes of property and power. As in earlier chapters, I include compass directions in my close description of landscape to orient the reader who chooses to follow the maps of the ditches.

Modern Minor Ditches on Private Plots of Land

A number of smaller ditches were built recently enough to provide access to direct observation by surviving informants. I examined these, looking for insights into how their builders designed them. In addition to the two main ditches currently in use, from the Qewar past Mallma Alta to Tantarcalla and from the Qewar through the Qolla Wayku gorge to the pampa, there are several smaller ditches. The largest of these waters Llulluchayoq, San Miguel's widest, flattest shelf on the slopes that lie above the ditch to the pampa. This ditch is said to have been built sometime after the turn of the century by Daniel Galdos' workers. It follows that path of the old high ditch, though, so it is in fact more accurate to say that Galdos directed his laborers to reestablish a section of the abandoned ditch. Contemporary Huanoquiteños do not recognize this, though. This shelf in 1984 still belonged almost entirely to the man who bought it from one of Daniel Galdos' heirs in 1968, who worked it with his sons and sons-in-law. This same man also owned the land over which the ditch flows. He and his sons raised the last part of the ditch's trajectory, and maintained it with the help of a few hired laborers (see E on map).

The Llulluchayoq ditch starts at Unuraqina, the confluence of smaller streams with the Qolla Wayku at the end of its upper valley, just before it enters its steep second gorge. A small short ditch, less than 100 meters long, diverts waters from a stream coming down the south-facing west end of Chuspi Grande to the ditch's mouth in order to increase its flow. From this point, at 3,661 masl., the ditch flows northeast for 782 meters along the lip of the Qolla Wayku's gorge, then curves to east-northeast, following San Miguel's slope as it opens away from the gorge.

The ditch starts into this curve at 3,638 masl., 117 meters above the ditch to the pampa's course across this same slope. The high ditch continues to parallel the lower one, but it has to cross the much steeper and more broken upper slopes of San Miguel, curving repeatedly from northeast to southeast and back to northeast as it follows the walls of deep gullies in and out of the mountain side. It flows over a long retaining wall, up to 15 meters high, and crosses log trestles in this difficult trajectory. Despite the unfavorable terrain, the walls of this privately maintained ditch are all in very good condition and do not show any of the signs of erosion and undermining we saw on the ditch to the pampa.

There are no outlets from this ditch in its first 1.5 kilometers, but then a series of small canals follow gullies down to scattered fields on small shelves, some of these quite steep, but less so than the general slopes, and on below these to the shelf already described above the ditch to the pampa.

The Llulluchayoq ditch now drops more rapidly than the lower one, partly because of the difficult terrain, and partly because there are no fields to irrigate close under the upper end of its course. One thousand seven hundred horizontal meters from its source, it has dropped 111 vertical meters, to 3,550 masl. It drops another 24 meters in the next 500 of run. Then it splits into two ditches, both of which drop about 30 meters, flowing fast, into small gullies and then over the edge of the much larger gully which runs northeast to Huanakauri. All this occurs within less than 2.5 kilometers from its source at Unuraqina.

When the family that farms Llulluchayoq is not using the water, they allow it to flow into the gully and then down it, where it joins the water in the ditch to the pampa under the waterfall at Huanakauri. Originally, instead of dropping into this gully, the ancient ditch turned sharply southwest into it for about 30 meters, then doubled back along its opposite side. From here it continued southeast at the top of Tanqarpata just at the base of the final steep rise to the summit of San Miguel. Only scattered traces of this ditch remain, all of them around the faces of small ridges, all of them disappearing into deeply carved gullies. The ditch continued down the top of Tanqarpata as it descends southeast of the pueblo, passing near a draw at about 3,200 masl. where some people told me the pre-colonial pueblo of Huanoquite was located. During its course across Tanqarpata, the old ditch passed both above and below extensive, now abandoned terraces, some apparently of Incaic construction, others considerably more recent, probably less than 100 years old.

Galdos may have resuscitated this section as well. I was told that this now abandoned section of ditch broke sometime between 1935 and 1940. Older informants have said that this land was still very productive in their youth; one of them said an earthquake had caused some of the sliding on Tanqarpata. The gullies into which ditch traces disappear have obviously deepened since the ditch was constructed, and it seems likely that the ditch itself contributed to this gullying. As I suggested earlier, ditches are weakest in gullies, and most vulnerable to destructive water flows there. Once a ditch has broken, it tends to channel water to the still intact sections of the ditch, which carry the now concentrated stream down to the break in the ditch. Thus, the water from a wide area is gathered into a single, narrow downward flow, deepening the gully even faster than rainfall over an unmodified slope would.

Tanqarpata is steeper, and its volcanic soils are looser, than other slopes on the same mountain that were formed as volcanic magma pushed up the predominantly sedimentary rock base north of Huanakauri. Nonetheless, the evidence of recent terracing, the fact that Galdos thought it was worthwhile to rebuild the old ditch there, and the reports of good production on this slope until 45 years ago all indicate that Tanqarpata was eroded to its present barren state in this century, and that irrigation aggravated the damaging effects of intense cultivation.

The ditch from Unuraqina to Llulluchayoq flows generally southeast, with the overall slope of San Miguel's north face above the pueblo. This ditch carries slightly less than half the volume of water that the ditch to the pampa carries. Another private ditch, much smaller (about 20 cm. wide by 10 cm. deep), but also about 2.5 km. long, flows generally north across near the top of San Miguel's east face to a point high above the abandoned section of the old ditch across Tanqarpata. This ditch flows against the overall southward slope of San Miguel. It was built about 40 years ago by a man who bought an eroded section of Tanqarpata from another Galdos heir (see F on map).

Starting from a spring that flows out of a hillside below and northeast of Yachipunku, at 3,772 masl., this ditch curves in and out of a series of sharp gullies and across the points of a number of small ridges, often just at the end of major drops into much more broken land, then turns briefly east in a long wash before dropping steeply north down to a small field above the old terraces in Tanqarpata. The builder of this ditch told me that it serves primarily to take water to his house, which is located just above its final drop, for him and his animals. He also uses it for very slow irrigation of the various small agricultural plots he has reclaimed from these rocky slopes. There are also a number of smaller ditches, also built by individuals, which distribute water from the multiple springs in and above the Qolla Wayku's upper valley. All of these are less than one kilometer long, and serve primarily to take water to houses and corrals, but they do also keep grass green along and at the end of their courses, and a few people in the sparsely settled valley use them to water house gardens as well.

The community of Maska tried to build a larger and longer ditch down this valley from the Qolla Wayku, to water a large flat plot near Unuraqina, but this ditch was never used, because the community never bought the tin sheets necessary to carry the water across a gully. There are two short ditches, one less than 100 meters long, the other a little over 200, that take water from the stream which runs down Ñekes' south base to small shelves on each side of this stream just above the point where it falls into the Qolla Wayku. There is also a small ditch that runs 650 meters along the right bank of the Watanay, starting just below the point where the old Inca road comes into the north and of the pueblo. This ditch waters several small fields before ending in a 2 hectare field overlooking the Qolla Wayku. It was constructed by the state officials who ran the salt mine just above the fields in which it ends. Another short ditch at the bottom of the Qolla Wayku irrigates three small fields, which total a little over 1 hectare, on the right bank of the Ccorca valley, under the northwest end of the pampa (G and I on map).

The high ditch to Llulluchayoq is the oldest, the largest, and the most elaborate of these minor ditches. It is the only one of them with high, long retaining walls and trestles, but most of these were rebuilt versions of the pre-Incaic original. Its original construction demanded technical skill and a great deal of labor. Even so, this ditch did not require the precise topographical understanding necessary to conceive other sections of the old high ditch, or even the two major ditches still in use. It flows around and down a single hillside, that is, the north slope of San Miguel from Unuraqina around to the highest of the shelves above the pueblo. Its source is sufficiently high that there was no need to keep the ditch bed at minimal drop in order to maximize irrigated area, so the calculation of level did not need to be particularly exact.

Of the remaining minor ditches, the 2.5 kilometer ditch that flows north against San Miguel's prevailing south slope is topographically the most complex. After discovering this ditch, and learning that it had been constructed by the man who still used it, I hoped that I could learn from him some of the ways that older ditch builders might have understood and conceived of the slope of hills in order to design their ditches. He told me, however, that a state surveyor, who had come to measure the boundaries of different private properties, had told him that it would be possible to take the ditch from its present source to a point near his house. He himself supervised the work of getting the ditch there, using the same technique that other builders of simple ditches described to me, letting the water loose as the ditch progresses and digging each following section the least amount necessary to allow the water to flow forward. He also said that the experience of directing irrigation water in fields made him aware of how to do this (H on map).

My examination of his ditch showed that he had to take multiple topographical features into account as the ditch wound in and out of the many gullies, washes, and ridges he had to take it around, and he had to choose between building more difficult retaining walls in order to stay high or dropping down to easier trajectories. In fact, however, he usually chose the easier solution, dropping the ditch across the most difficult traverses, an option which was viable because the source was so much higher than his destination that he did not need to build for minimum drop. Once he had been told that the ditch was possible, there was no point at which any of his decisions obliged him to take into account more topographical features than could be seen simultaneously from the same point.

The man who built the short ditch from the Qewar told me rather vaguely of a system of posts which could be used to sight the drops, but he described this procedure in terms of a single sight line, and he did not mention any device for measuring the kinds of exact level which minimal drop would require. Nonetheless, it is conceivable that series of precisely measured posts of the same height could be placed in a sequence if there was some way of sighting across successive posts so as to position them along the minimal necessary drop. These posts could be set in sequence at the proximities necessary to allow a continuous sight line between succeeding posts, so that the trace could be followed around the curves in hillsides. This procedure, something like a series of stadia boards, would still require a precise leveling device to assure minimum fall. Working as a surveyor, and with constant practice, I had learned to estimate fairly precisely what exact level would be, even when standing on steep slopes which at first had deceived the eye. I do not believe, however, that even the best trained eye could achieve the precision necessary for the minimal drops, less than 1 in 100, which I saw. I still do not know, however, what kind of more precise leveling device the ditch builders may have had. Nor can I know, therefore, whether the builders of the ditches still in use used the same devices and techniques which the builders of the much older, now abandoned ones did. It is possible, as I mentioned before, that the more recent ditches followed the traces of older ditches. This would be possible, though, only if the old ditches were sufficiently well built and well maintained that they did not eventually ruin their own trajectory, through erosion.

It is also possible that ditches were designed, and their construction overseen, by specialists. The assumption in some historical accounts that the Inca state engendered a series of specialized social groups suggests this as quite a strong possibility for the oldest ditches. The evidence I presented in the last chapter indicates, though, that the topographic knowledge required for the old high ditch would be accessible to long, attentive experience in and of the local landscape. Itinerant specialists would not have the necessary knowledge. Nor would the Incan empire have had the logistical capacity (cf. Mann, 1982) to organize and coerce the labor necessary, especially with the much smaller population that Huanoquite could sustain before the ditches were built.

I found tremendous variation in both the extent and the accuracy of geographical information as well as in skill in building among the present day Huanoquiteños. There is no reason to assume that topographical understanding and technical skills were any more equally distributed under earlier systems. The difference between then and now, though, might well be that under the more decentralized power and authority structures of both the pre-Inca community and the Inca state, local people with special knowledge would be accorded consensual authority over such projects and of the large labor force necessary to carry them out. In contrast, the division of land and labor into separately owned coercive units under the hacienda regime, and the present, far more dispersed system of authority within community and cooperative, lack any clear coordinating authority. Neither post-Inca system could engender the consensual solidarity, necessary to implement any conception, vision, or technical skills needed to build the original systems. As a result, even ditch repair is hard to achieve, and major new projects almost impossible. The heavy time pressures which a) recent diversification of production, b) the access to greater amounts of land, c) the need each farmer now has to make his own arrangements for transport and sale of crops and to acquire for himself whatever fertilizers he may use—tasks he did not have to undertake when most of what he produced belonged to the *hacendado*, and d) the increasing number of private plots of land, makes individuals, and therefore collectivities, reluctant to undertake major new projects. Under the hacienda regime, only the exceptionally charismatic Daniel Galdos was able to coordinate the different autonomous owners' control over labor. His powerful successors to mayoral authority were not able to coordinate adequate maintenance, let alone to inaugurate new construction.

Even the maintenance of existing infrastructure and participation in small scale new projects already puts a heavy burden of *faena* labor on both the community and the cooperative, which individual members have to fit in with their own labor requirements. The unequal access to water which results from continuing inequality of land tenure and land quality increases this reluctance to dedicate more labor to ditch construction or repair. The community's failure to complete the simple ditch its members started in the Qolla Wayku valley may be symptomatic of all of these problems.

This would also explain why only a few individuals, including the builder of the small ditch which runs north across San Miguel's south slope, have expressed concern about the deterioration of the major ditch from the Qewar to the pampa, and have repeatedly proposed solutions which would require far more labor and material than are presently dedicated to the annual ditch cleaning *faena*. Most of the people I talked to were either unaware of or unconcerned about the structural decay of the ditch, or simply said that it is a matter for the state to attend to. In contrast with the few people who expressed concern about the condition of the ditch and urged major work on its reconstruction, the majority show less knowledge about the ditch itself, about the effects of continued erosion on it, about the way it was built, and about what kinds of repairs or even change might make it less likely to collapse. This lack of interest and knowledge tends to be associated with, obviously, a reluctance, or simply lack of response, in the face of suggestions that more work needs to be done on the ditches. Thus, the idea continues that the ditches are received, a gift from the past which imposes some inconvenient labor demands but which generally can be considered a permanent and unchanging part of the environment they use, and that if they consider their construction at all, it is in the sense that the people of the past were better, stronger, and harder working than they are.

The Mills

In addition to irrigation, some of the newer ditches served power mills for Huanoquite's large wheat crop. Two of these were located on the left bank of the Qewar's upper valley, about 270 and 390 meters respectively from its source. A ditch about 30 cm. deep and 30 cm. wide, lined with stone and a limestone cement, led water up the side of the bank from just below the Qewar's source, then dropped it down a narrow chute between 1.5 meter-high walls, and into a hold halfway up the back wall of a 2.2-meter high, 4-meter long, cemented stone arch. The force of the water turned a large upright beam, anchored on the floor in another beam which went across the arch and at the top in a hole carved in a large stone set at the top and middle of the arch. The upright beam turned a grindstone 1.4 meters in diameter. The arch and beams are still intact, though the grind-stone now lies outside the arch, and what appear to have been other buildings have all fallen down. The entire structure is overgrown with dense bushes. A ditch to the second mill, which is identical in construction but did not have any other buildings, starts just below the first.

These mills were both constructed under the orders of Cirilo Paz, who sold the wheat ground here as far away as Cuzco. They have apparently been abandoned for over 60 years. A third mill originally owned by Paz still stands just below the Qewar's pass out of its main valley between Chuspi Grande and Wataqasa. A ditch takes water from the Qewar, around a steep 85 degree crag, and through a steep chute under this mill, which functioned until fairly recently, but was badly maintained and so worked at a very low volume. There are still several adobe ovens in Huanoquite, but people are now obliged to take their wheat to Cuzco, have it ground there, and then bring it back to Huanoquite if they wish to bake bread. Like the private ditches, the mills show ingenious adaptations to the topography and hydrology of a small, privately owned space, but require little sense of the multiple systems that configure in an integrated landscape.

The Terraces

The association between the decay of deliberate environmental modifications and the loss of technical and historical knowledge is even clearer in both the condition of, and what people say about, the extensive terraces on many of the slopes in Huanoquite's major agricultural areas. Terraces, or traces of their remains, are still visible on hills with a variety of different slopes and elevations. Some, like the scarcely visible traces above and below the abandoned ditch on Chuspi Grande's north face, or those which have survived much better on Parapara, are on steep slopes, up to 30 degrees. The terraces on Ñekes cover slopes which reach 38 degrees. Others, like the sod walls which fit Urubamba's two-way slopes, or like the rock walls which form the seven large terraces just below Paqpiri, break up gentler slopes, between 5 and 11 degrees.

The height of the walls varies both with the width of the terrace and with the slope of the hill. Some on Parapara are 3 meters tall, while those supporting narrow strips on Urubamba reach less than half a meter. Some of the terraces are large and continuous, like those below Paqpiri and those on Parapara, others, like those on the steep drop east of Urubamba, support tiny shelves scattered across the parts of the slope where a small break in the overall drop could be accentuated and fortified. Some are no more than rows of sod which supported the lower end of fields, allowing a row of permanent vegetation to grow up and catch some of the soil loosened by cultivation at the top end of the field, thus slowly leveling off the entire field while increasing the height of the terrace wall. In others, maguey has been placed along the wall, both to fortify it and to keep out animals who would both graze on the crops and break down the walls. Sometimes the rocks removed from the cultivated areas have also been placed along the tops of these walls.

Other walls, in Saqsi, in Parapara, below Paqpiri, on Ñekes, and on parts of Tanqarpata are built of rock, some large and carefully fitted, others smaller and less sturdily put together. The walls above Paqpiri and the long terraces at Saqsi are built of large heavy rock. These are associated with other structures, perhaps watch towers, which suggest Incaic construction. Some of the walls on Ñekes appear to have been made with much smaller stone, and may be more recent. High on Tanqarpata are long walls of round volcanic stone, a precarious building material now mostly tumbled down, I was told were built when Daniel Galdos owned that land. Lower down on the same slope, near Huanakauri, however, much more elaborately constructed terraces circle a cone-shaped hill, which is topped by what also may have been a watch tower and is certainly much older than the terraces higher up.

Many of these terrace walls are quite obvious, others are now barely discernible, and about some I am simply not sure. I have seen long, straight breaks across hillsides, from half a meter to just under a meter in height, which suggest that these areas were terraced, but there are no longer rock or sod remains to confirm this. On many of the steeper hills, there are numerous small flat areas with severely eroded down sides. These have been and still are cultivated. Again, on many of these, there are no evident remains of retaining walls. Given the severity of the erosion under these areas, and the stark contrast between their flat tops and nearly vertical edges, it is hard to imagine how these would have occurred naturally, or how they could have been cultivated for as long as they have been without retaining walls below. Indeed, the frequency of flat areas as breaks in steep slopes throughout Huanoquite's agricultural area strongly suggests long-term, repeated modifications of the environment, which may only have been allowed to deteriorate in the recent past. This speculation fits with the evidence of extensive long term occupation and production suggested by the extensive ditch system itself, the various watch towers scattered about over a wide area, the major Incaic highway passing through Huanoquite (perhaps in part because of the salt which could be mined here). Incaic remains on top of Llamapintay, and the carved thrones located where the old road enters the pueblo. All of these indicate that Huanoquite has long been an important center of intense agricultural activity.

If this speculation is correct, the terraces contributed importantly to agricultural production and protected the soils against the pressures of agricultural intensity. Most of these terraces have been neglected, and those that remain are conserved either by the accidents of private property ownership or for reasons other than erosion control. The terrace walls on Ñekes are now mostly rock piles at the bottom of the slopes, though middle-aged men remember them from their childhood. Ñekes' steep slopes are no longer cultivated; the terrace platforms are still visible, but are already returning toward the general slope of the hill. Cattle and horses graze there now, and their weight may have contributed to the walls' collapse. The terraces above Paqpiri, on flatter ground under the end of Ñekes, are still more or less in place, but these too are starting to crumble. Old apple trees, no longer producing, indicate that this area was cultivated in the fairly recent past, but there is only grass here now, and grazing cattle accelerate the destruction of these terraces as well. One of several claimants to this part of Cirilo Mora's old hacienda told me that this area is no longer cultivated because its inheritance is still at issue, and other informants have explained that there is a conflict here between legitimate descendants of Paz's daughter and the heirs of a son she bore before marriage. Nonetheless, terraces are neglected even in areas where there is no question or conflict over ownership.

Some of the high rock walls at Parapara have also broken down, but here, dense vegetation along many of the walls protects them from animals and fortifies them against erosion. Community members continue to cultivate these plots, so they are interested in keeping animals out, but some of them do burn the vegetation to get rid of birds which they say eat the seeds they plant there.

Some of the people I talked to expressed concern with erosion, but almost always in response to leading questions. The issue seldom arose spontaneously, even in conversations where it would have been quite relevant, and many of the people I talked to denied it was a problem. Most people are aware that the old terraces served to prevent erosion, but even the best informed appear to identify as terrace walls only the obvious rock constructions like those below Paqpiri and on Saqsi. The rest are most commonly described as *linderos*, boundaries between private plots. One man who lived in the upper Qolla Wayku valley, and who knows the area well, even described the rock walls on Parapara in this way, and specifically denied that they served any function of erosion control. On another occasion, I stood talking to a man on the ditch above Urubamba, looking down at the terraces from a point where their perfect fit along the different contours of both the pampa rim and of one of the depressions through Urubamba could be precisely seen. This man as well answered my deliberately naive questions by describing them as *linderos*, and answered my more specific question by saying no, they did not serve as terraces. When I pushed, asking why then did people bother to maintain them so carefully, he said it was to keep their neighbors from encroaching on their plots.

If he was correct in this, then the protection of private property does serve the accidental function of maintaining the terraces and preventing erosion where land is divided in small parcels. It may also mean that successive fragmentation tends to follow boundaries already marked by terraces, and this too would tend to preserve the terraces. These functions do not, however, work for land held in large plots, or for land held in common by the community and the cooperative, unless, as seems to be the case on the community lands in Parapara, the same family continues to have usufruct of the same land for a long time and even across generations.

Thus, the terraces continue to deteriorate unless they serve to mark private property. Even more dramatically than the ditches, they are an environmental modification received from the past which importantly affects production capacity in ways which present-day Huanoquiteños take for granted, with little concern for their maintenance; even the knowledge of the purposes or functions of some of these terraces has been lost. Again, part of the problem seems to be the pressure of time; when I asked one man how he controlled the erosion on his land, he answered "No nos alcanza el tiempo," there is not enough time. Some terraces were still maintained, and some new ones built, under the haciendas, but the remaining large land owners no longer have the control of labor, and the new small holders are too busy with the daily and yearly round of agricultural tasks imposed by their different crops and by the care of the animals many of them now have, to maintain the terraces or to build new ones.

Most of Huanoquite's farmers appear to be more concerned with the crops they grow than with the land on which they grow them. They are aware of erosion, and water more slowly on steep grades, but deny erosion as a problem in the fields they are irrigating, even when I stand with them and watch irrigation water furrowing the ground before they have planted it. Erosion from irrigation or from rainfall is discussed as a problem more in terms of its carrying away seeds and fertilizers after planting; this is a disaster people worry about and guard against. Plowing and planting patterns also take the crop rather than the ground into account. Potatoes, for example, are planted in rows running downhill in order to assure that water does not stay in the ground and rot the plant. Corn is sometimes planted in contoured rows, but more frequently plowing is done in such a way as to maximize drainage, though the rows of corn are not as deeply furrowed as those of potatoes, and runoff does not cause as much erosion as with potatoes. When the rains come at the end of the planting season, the rivers, which were clear during the preceding months, turn deep brown, carrying off soil which will renew the fertility of the Amazon's old and geologically impoverished food plain at the expense of long-term use on Huanoquite slopes.

This erosion is exacerbated by extensive deforestation, which may have started early in the colonial period, when the Spaniards occupying Cuzco demanded vast quantities of lumber and firewood from the surrounding areas (John Rowe, personal communication, 1984), and certainly continued under the haciendas, when the *hacendados* sent wood from Huanoquite both to Cuzco and to other provinces. Huanoquiteños continue to chop down trees for building, for firewood, and for agricultural implements such as plows. Some farmers have planted eucalyptus, both for consumption and for erosion control on the high slopes of hills, but various of my informants told me that they remembered hillsides with many more trees than are there now.

Roads Old and New

Another environmental modification has continued from the past to the present in different forms that reflect radical changes in technology and in political systems. The Incaic road from Cuzco continued on past Huanoquite across the Apurimac river on the Huacachaca suspension bridge and from there perhaps on to Arequipa. Huanoquite is also connected to the surrounding communities and *anexos* by a network of footpaths, which the Huanoquiteños call *caminos de herradura*. Like the ditches, although in less elaborate ways, these roads must take topography into account. Far less than would occur on flatter terrain, there are entirely separate routes between Huanoquite and a series of outlying points. That is, because the roads tend to follow ridges and valleys, they tend not to bifurcate from a central trunk, but rather each to have their own best trajectory. Some of these are simple paths, worn clear by many feet over the years. Others, however, including the Inca road, travel along routes which have been built up on rock walls across gullies, or have actually been cut into hillsides. These legacies of Incaic and pre-Incaic effort became the responsibility of the district Consejo, and *faena* labor was organized to build and maintain them. The road to Cuzco required coordinated work by several districts. These remain in use, and are clearly very important in travel and commerce between the different communities and *anexos*. The cobbling of the major roads, however, some of which were still being maintained early in the last century, is no longer kept up, and is now deteriorating.

The old Incaic road leaves Huanoquite's northeast end, past the throne carved in the limestone overlooking the Watanay. It travels down to the Ccorca, first along the Watanay's, and then along the Qolla Wayku's, right bank. After crossing the Ccorca, it rises steeply along a path cut into a hillside of the Ccorca valley, then turns sharply into the Miska Wayku gorge, along a cut halfway up this slope. Large stone steps have been set in all of the steep places. The road continues due north to the narrow pass at the top of Miska Wayku, then turns northeast through a long wide valley, then north through another valley, Chercoperca, before finally crossing Quipan—the sacred site at 4,000 masl. overlooking Cuzco, 720 meters below and 9 kilometers away—and then continuing down.

In 1964, the *alcalde,* or mayor of the district, successfully solicited state funds for machinery, materials, and *faena* labor to open the truck road, now in use to Yaurisque, where it connects with another rustic road from Cuzco to Paruro, the provincial capital. Trucks largely replaced horse and mule trains over the old Inca road to Cuzco, but the long detour through Yaurisque motivated several attempts to build a direct road to Cuzco, along a route close to that of the Inca road. All of these attempts failed, primarily because the state withdrew support. Another state-financed road from Huanoquite to Huanca Huanca, near the Apurimac, got as far as the Abra de Vilcacunca, but stopped there, again for lack of funds. Thus, Huanoquiteños continue to use the long route around, or walk. The truck ride is long and uncertain, trucks leave and arrive irregularly, break down or get stuck, and sometimes have terrible accidents on the sharp grades and steep drops. Because most individual farmers sell their own crops and buy supplies in Cuzco, however, the trucks are essential. People agree that walking is a surer way of arriving, but as their trips usually involve buying and selling, most take the trucks. Roads require a far less intimate mediation between society and nature than ditches do, and so do not directly inform us about environmental sensitivities. On the other hand, the steady shift of responsibility from local to far less predictable national jurisdiction is as relevant to ditches as to roads.

Environmental Modifications, Past and Present

The ditches, the terraces, and the roads have all played an important role in shaping Huanoquite's landscape, in making parts of it cultivable and then more productive, in making possible the exchange of goods with nearby communities as far away as Cuzco, and thus in supporting a larger population than would otherwise have been possible. The breakdown or deterioration of parts of these infrastructural devices, on the other hand, has reduced the usefulness of some parts of the environment. Political and economic systems have changed drastically since the Inca conquest. Huanoquite continued as an important agricultural center within the Inca Empire (Seligmann, 1987). After the Spanish conquest, much of the land in the area was controlled by different religious orders, and by the 17th century haciendas had taken control of most land and labor. *Hacendados* also controlled the district Consejo, which had a right to demand *faena* labor and which also was in charge of many of the infrastructural projects, including ditches and roads. Though the community of Maska was recognized long before the agrarian reform, it had very little land until it won several cases against some of the large land owners in the 1970s. Peasants not only gained access to plots within the community holdings, but some of them also started to buy small pieces, often from the heirs of the *hacendados*. They now farm more land, and they farm it harder.

The Agrarian Reform brought about important changes in the control of land, the size of holdings, the intensity and kinds of cultivation, and the variety of tasks that the individual farmers were obliged to attend to. Farmers now have to sell their products in Cuzco, and also have to travel to arrange credit, buy fertilizer, and, for some of them, to participate in the lengthy court cases over land reform or to attend to the business of the community and the cooperative. They are now in control of their own lands, and free of the obligation to work for the *hacendados*, but they must use the time they now control for a much wider range of work. They continue as well to owe *faena* labor, to the cooperative, to the community, and to the Consejo.

The demands for *faena* labor have also become more complex as life styles and economy have changed. It is essential to maintain the truck road to Cuzco. This road is longer, wider, and more susceptible to erosion than the old road was; trucks get stuck more easily than mules, and damage the road more. There is more demand for education, so schools must be built and maintained, as must a health post and a building for the Consejo. *Faena* labor is also required on the community and cooperative fields which generate income for these organizations.

The structure of political power and authority also changed drastically. Prior to the reform, the various *hacendados* fought, sometimes violently, among themselves over the control of land, the control of labor, and the control of formal political office. Their control of land and labor was not absolute; indigenous communities continued to have formal rights, and some of them remained free of hacienda control. After the 1930s, there were official government agencies responsible for their protection.

Nonetheless, within Huanoquite itself, the hacienda system was nearly hegemonic. Even in terms of formal rights, the communities were weak, and what formal rights they did have were frequently violated with impunity. *Hacendados* controlled the land, the Consejo, the courts, and the police. They used this control to exact hard labor and subservience from most of the Huanoquiteños, and it was only the exceptional agent responsible for the welfare of the communities who was willing to challenge them seriously. Whatever their personal rivalries and conflicts may have been, they were fundamentally interested in maintaining the system that gave them power, and in exploiting the land and labor they controlled. Thus, the construction and maintenance of infrastructure involved agreement and coordination among a relatively small number of people. The ditch from the Qewar to Tihuicte passed through only three major holdings, all haciendas, at the beginning of the century, and the ditch from the Qewar above Mallma Alta to Tantarcalla passed over one hacienda before reaching the community. The men who owned these lands also controlled the district Consejo, so they had little difficulty in mobilizing the labor necessary to build and maintain the ditches.

The situation today is very different. The ditch to the pampa now crosses many different properties, and the political authority to maintain the ditch is now fragmented between the community of Maska, the Tihuicte cooperative, and the district Consejo. All of these separate authorities must respond to their numerous members, and all of them compete over authority and power, as well as over access to material resources and to labor.

The fragmentation of land holdings, the liberation from peonage, the diversification and acceleration of individual and household agricultural tasks and of collective projects, conflict between different social groupings of land holders, and the dispersion of political authority among three partially opposed and competing entities all tend to impede coordinated effort for the initiation or maintenance of environmental modifications and infrastructure. The natural processes of deterioration are thus not as effectively retarded as they previously were, and new areas are opened to intensive cultivation without the safeguards for protecting the soil which previously existed. Such deterioration is not new; the abandonment of the old ditches, and the erosion visible along some of their trajectories, indicate that there may have been other periods of infrastructural collapse, and perhaps as well of agricultural crisis.

The old ditches stopped functioning before there were alternative irrigable areas under other ditches, but I cannot know how long a period might have passed without irrigation. It is possible that a period of social and political disorganization followed the Spanish conquest. Severe depopulation after the Spanish conquest, also a strong possibility, could have further impeded ditch maintenance at the same time that it reduced the need for irrigable areas. As I cannot precisely date either the construction or the collapse of the old ditches, however, I have no way of doing more than considering these possibilities in the light of what the ditches and the areas they once irrigated now look like.

I can, though, tell much more about the current deterioration of the ditches and terraces, about the difficulties of organizing labor to do more than make the quick repairs necessary to get through one more dry season, and about the virtual impossibility of organizing labor to build new ditches or terraces. All of these are pressing needs; the collapse of existing ditches would severely and immediately reduce both commercial and subsistence agricultural production and the continued collapse of terraces is slowly reducing the land's productivity. There is already great pressure on the land, and irrigated land is much more intensely used than rain-fed lands; both because it is more productive and because it can be planted before the rains begin. To compensate for the unequal distribution of water, poor families can sharecrop or provide labor for more prosperous neighbors. Without the ditches, the households which are barely producing enough for their own livelihood, probably the majority, would face nutritional deprivation, and the smaller number of households which are expanding their agricultural production, acquiring more livestock, building new houses, educating children, and increasing their stock of tools would be reduced to a hard scrabble just to feed themselves. Thus, the collapse of a major ditch would have dire consequences for many families.

The building of new ditches is not as calamitously important as the maintenance of old ones, but it could alleviate the growing pressure on the land and reduce the inequalities in access to productive land. The community of Maska has greatly increased its land holdings since the reform, but very little of the community's land and none of what it farms collectively or distributes among its members is now irrigated. Thus, both corn and early (*maway*) potatoes can be grown only by private holders with land under the ditches, by members of the cooperative, either collectively or on the personal plots the cooperative provides to its members, or by individuals who enter into sharecropping or rental arrangements with private holders. Sharecropping and labor rental are not favorable options for most farmers, but are the only ones available to the poor, who do not have the money for monetary rent. Labor rental requires about 1 day of work per week for a whole year in exchange for about one third of a hectare; the sharecropper provides all the seed, the fertilizer, and most of the labor in return for half the crop. This means, among other things, that he has to more than double return for any investment he makes on fertilizer, an unlikely outcome given the high price of fertilizer and the low prices of farm products. For these reasons, building new ditches to community lands would significantly improve the economic opportunities of many community members; so the community's inability to organize major infrastructural projects reduces the potential welfare of most of its members.

I discussed several possible explanations of the community's, the cooperative's, and the Consejo's inability to organize the labor and materials necessary to maintain adequately existing ditches and terraces or to construct new ones, either separately or in conjunction. These various explanations are mutually compatible, and the factors in them are mutually reinforcing. Fonseca (1984) however, has described the social and political organization of corporate groups which do, apparently successfully, maintain irrigation systems in the Cañete valley, and traces the system of community control which does this back to the Inca state. Presumably, the members of the communities that Fonseca describes suffer many of the time and labor pressures that affect Huanoquiteños. The important difference between them may be, therefore, in the history of political arrangements. Haciendas were established later, and were never as powerful in Cañete as they were in Huanoquite. Thus, communities had more economic and political function and cohesion there.

Many Peruvian communities have been divided by the incomplete or partial effects of the agrarian reform, which included some families in cooperatives, and left others out. Thus the division between community and cooperative in Huanoquite is not unique. Nor, however, is it universal in Peruvian farming communities. The history of nearly hegemonic hacienda control is also far from unique to Huanoquite, but neither is it universal in the Peruvian highlands. Huanoquite suffered both tight hacienda control and then an incomplete agrarian reform which created clear differences in its inhabitants' access to land, technical assistance, and material resources, as well as in the types of collective association which distributed and to a certain extent controlled these components of agricultural production.

Fonseca's data directly invalidate any direct and simple application of Wittfogel's (1957) thesis that irrigation civilizations require centralized authority. Wittfogel himself, however, had clear difficulties in fitting egalitarian systems like the Southwest American pueblos into his model. Wittfogel's discussion of pueblo irrigation, and Fonseca's of the effectiveness of ditch committees in the Cañete valley, suggest that egalitarian communities can develop and maintain the elaborate environmental modifications which ditch systems require. Reichel-Dolmatoff (1976), writing about a very different kind of society and economy, has shown how communities of jungle Indians develop the kind of belief systems and social control necessary to impose collective interest in environmental preservation over individual interests in maximizing returns on individual effort. Such systems, and this kind of social control, require a highly integrated, well organized community. It is not surprising that Huanoquite has not been able to develop this, given the tight *hacendado* control which persisted until a decade and a half before my field work, and is still not completely broken, or given the way that the Agrarian Reform's unequal effects divided the inhabitants of the pueblo even more recently. There is another factor, however, which may impede the construction of new ditches. Most Huanoquiteños did not appear to have a sufficiently clear notion of the topography around them to design ditches. For many, the idea of the landscape appears to be as fragmented as their own parcels of land are and as the numerous place names for very small sections of land would suggest. They may know well the agricultural characteristics of each small parcel, of its soils, climate, etc., but the overarching conception of the relation between different valleys and slopes which would be necessary to design ditches is not at all apparent, even in conversations specifically directed to questions of how the old ditches were built and where new ones might go. This was especially true in most of the conversations we had about the now abandoned ditches and why they might have ceased to function.

Attitudes and beliefs about the old ditches are extremely important, because these ditches watered the areas which are now some of the community's most important lands—Chuspi Grande and the slopes and floor of the Qolla Wayku's upper valley above Unuraqina. Any attempt to provide significant amounts of irrigated land to community members would involve either resuscitating these ditches or building courses similar or parallel to the old ones. A few community members, especially the president, became interested in this possibility when Plan Meris, a joint German-Peruvian government program for building irrigation ditches, chose Huanoquite as the possible site for one of their projects. Plan Meris, however, stipulated certain minimum criteria for the volume of water available for any ditch it builds, and maximal criteria for the amount of slope below its ditches. The Qewar has less than half of this minimum volume, and the lower ditch across Chuspi Grande crosses above ground much steeper than the 15 percent grade specified by Plan Meris. The community's president was particularly interested in resuscitating the upper ditch which would water the relatively gentle western slopes of Chuspi Grande, and the wide, flat, fertile expanse of the Qolla Wayku valley.

Only very few members of the community, however, seemed particularly concerned with the possibility of opening the old ditches, and even these seem to think of it as a project for Plan Meris rather than as something the community could undertake. One of the most important reasons for this may lie in the explanations they gave me for why the old ditches ceased to function and the way these explanations fit into a complex set of beliefs about the magic of water and the powers of the Incas over it, and the way these beliefs affect present-day Huanoquiteños' understanding of their environment.

I had first become aware of the old ditches when I went with a farmer to look at his wheat atop Chuspi Grande. I knew that there was no water source there, and we had seen that Chuspi Grande was a large round mound, with slopes dropping off on all sides. Thus, I could not imagine where the ditch water had come from. I followed the ditch traces backwards, from east to west above Chuspi Grande's north face, but the going was difficult enough, and I lost the traces so often, that I did not get all the way to the end that day. I asked another farmer I met near my stopping point how the water had crossed over, and he said that there had been a large stone aqueduct built by the Incas to bring the water across from the ditches I could see coming down from Parapara.

The next day I returned, and discovered that no aqueduct would have been necessary at all, but that the land bridge formed by Pukaqocha between Chuspi Grande and Parapara was high enough for the ditch to cross over. This experience gave me two important lessons; first, that my own eyes had deceived me into thinking of Chuspi Grande as a mound with a flat top, which is what it seems to be when one walks along or under it, instead of showing what it really is, a large mound with slopes all around it, but which slopes sufficiently to the east so that the bottom of its west face, where it meets Pukaqocha, is actually higher than the top of its east face. Even knowing this to be the case, if I stood on Wataqasa and looked across at Chuspi Grande, it still looked much more even in elevation than in fact it is. Only from the other side, and much further away, atop Wankankalle, is it easy to see the true slope, but from the Wankankalle, one is standing on the opposite side from the ditch's course. In addition, one seldom has reason to stand on the point on Wataqasa which would most facilitate understanding Chuspi Grande's true shape; it was much more common for me, and for the Huanoquiteños, to see Chuspi Grande from other angles, and against the background of other hills, or from too close up, to be able to follow the line from the bottom of the west slope to the top of the east slope and see that it was nearly level but slightly downward.

The second lesson was that the farmer who told me about the land bridge had been just as badly deceived as I was, even though he had spent all of his long life in Huanoquite and knew the area very well. This experience broke me out of my unconscious, and very naive, assumption that all local informants knew everything there was to know about their irrigation systems and the topography they crossed; it helped me to start looking at the Huanoquiteños' geographical and topographic knowledge as individually and historically variable. The causes and effects of this knowledge, then, could tell me a lot about what possibilities individuals, and communities, might or might not perceive in their own environment.

As I started to follow the old ditches, I asked people about their routes and about why they had been abandoned. Everyone agreed about the route of the ditch around Chuspi Grande, but there were several different reports about where it had ended; some said at Unuraqina, others said that it had continued into the Qolla Wayku valley, and only a few realized that after going down into the Qolla Wayku, it had continued back across San Miguel's north slope to the fields above the pueblo. There was even more variety in what I was told about where the upper ditch had ended: several informants suggested that it had gone to the top of Llamapintay, to the site of Inkaraqay, which is much higher than the ditch could have gone. The answers I got about why the ditches had been abandoned, however, were more consistent, although some of the informants who told us expressed skepticism in the old stories.

The most common explanation I heard was that after the ditch was built, a group of men went to the source of the Qewar to open the ditch. A large snake with golden braids emerged, and led the water down the ditch. One of the men killed the snake, and the water returned to its source. Another version recounts that the animal killed was a frog, but both of these stories conclude that the Qewar then started to emerge from a source lower down the hill, below the opening to the ditch, so there was no way of getting water to the ditch. People who expressed skepticism about the snake, or about the idea that the ditches were ever used, do believe that the Qewar now emerges from a point lower than its original source and lower than the ditch outlets.

A variant of the same story explicitly links the Spanish conquest to their interference in both the practices and the meanings that grew around Huanoquiteños' relations to the ditches, the snake, the frog, the duck, and other animals. The variant version explains that the old high ditch is no longer used because when the Spaniards arrived and the cock crowed, the Huanoquiteños ceased making the necessary offerings to the snake with the golden braids. I believe that these offerings refer directly to the ongoing work of maintaining the ditches, and to the authority and respect accorded to the leaders most sensitive to their relation to the landscape. As I listened to stories about specific places and about the animals that guarded them from human abuse, I more and more clearly saw the snake as the embodiment of the relationship both between the community that cultivated and the earth that brought it the water to do so, and within the community of respect and reverence for those who regulated this essential relationship.

The rest of the variant story is the same. When the Huanoquiteños stopped making their offerings, the snake had cursed them, moving the gushing spring that watered the ditch further down the hill, below where the ditch could receive it. The snake then said that anyone who tried to restore the old ditch would die. This story did not mention the obvious fact that the ditch was broken and had caused serious erosion downhill from the breaks.

I had heard these stories while trying to figure out the courses of the old ditches. The job of locating and following the traces took long enough so that I had heard them several times, and in several different versions. While skeptical about large snakes at that elevation, I had assumed that there was no longer any way to get water to the ditches. On the last day of searching, knowing that the Qewar's source was just around the hill, I followed the traces to a knoll overlooking the river's drop through the trees to its lower valley, and suddenly found that there was water running through the ditch. The ditch was very easy to follow from there to its source, but I went back over this section to make sure that there was a constant fall between the part in use and the traces which were empty. I confirmed that indeed there was a constant fall, and thus that the Qewar's waters still emerged high enough to fill the old ditches. I was amazed that no one had told me about this short section of resuscitated ditch, and that no one I had talked to seemed to realize that its existence made impossible their explanations of why the ditch had been abandoned.

The ditch waters a small part of the lands belonging to one of Cirilo Paz's heirs, a man who had withdrawn from the pueblo to live in this removed site above the Qewar, and did not get on well with most of the pueblo's inhabitants. I wondered if he had perhaps managed to keep his ditch secret, or if people went to his land so seldom that they were unaware that he had opened it. In fact, however, the footpath to another community passes nearby, and it is quite obvious that parts of his land are being irrigated. It was only when I went to talk to him directly that I learned how powerful the beliefs about water, and about the Qewar, are, and why no one seemed to have seen the contradiction between the water in the ditch and their explanations about its abandonment.

He told me that he and his sons had opened the ditch in a month of very hard work, four years before, shortly after they had moved there to live. He said that they had searched for other water sources first, in places higher up the hill where they could hear the sound of wind and water in hollows below the earth. He told of finding one source and starting to dig a small canal from it, and of going back the next day to find that the water had disappeared. When he dug down where the water had been, one of his sons started to sink into a bog just below the crust of the earth. It was then that they started to work on the old ditch. He told me that the old men from the nearby community of Marcura had told him that he could never complete the ditch, that after the snake had been killed, the old people had commanded large stones to block the ditch's course, and then put a curse on the stones so that the arms of anyone trying to remove them would become numb and weak until they stopped trying. He said though that the curse was old, and had finally gone away, so that he was able to move some of the rocks. He insisted, however, that the old ditch course was still powerfully cursed and protected, and that he had had to follow a lower course.

I went back again, checked all of my measurements, and searched assiduously for other ditch traces higher up until I was convinced that, not only were there no other traces, but that there was no other possible course for the old ditch than the one which this man and his sons had reopened. He himself, though, apparently could not overcome his own belief that the ditch was cursed or that the Qewar now came out below the level that it previously had, and so convinced himself that he was following a lower course. In many subsequent conversations with other Huanoquiteños, only the president of the community, who had walked this ditch with Plan Meris engineers, seemed to know or believe that the old ditch could still be filled by the Qewar. The president told me that he had learned this from the engineers, and had been surprised to find the old story was not true, but even he thinks there was an even higher ditch before the water dropped.

None of this, of course, helps explain why the ditches were abandoned in the first place or where the belief about the snake and the lowering of the Qewar's source came from. After the ditch cleaning *faena* was completed, I climbed up the face of San Miguel to look down at the trestle which carries the ditch to the pampa across the Qolla Wayku gorge. A group of men had been sent back to open the ditch at the Qewar. As we watched, the water came around the bend into the gorge. It was carrying the debris—dirt and twigs and leaves—which the cleaning and repair had loosened, and the force of the water running across and falling over dry earth and rock had mixed the dirty water with air to create yellow and white spume. The water wiggled back and forth around the curves in the ditch, moving and looking from my high vantage point like a large brown snake. The spume trailed back along the top of the water, forming the snake's golden braids. Thus, I found the snake that led the water down the ditch, but I didn't yet know why or how the story that someone killed it came to explain why the old ditches were abandoned, or the mistaken idea that the Qewar started to emerge from a lower source after that happened.

It was uncertain whether Plan Meris would proceed with its projects in Huanoquite, which were then at the stage of feasibility studies. It was also unclear whether they could help the pueblo organize to maintain the new ditches while continuing to use the old ones. Plan Meris' proposal was to bring a large ditch from the Ccorca, far to the north, high above Tantarcalla, and then on across San Miguel. Some of this water would be released down to Tantarcalla, thus obviating its inhabitants' present susceptibility to Mallma Alta's owners' taking all the water from the present ditch before it reaches Wiñay Poco. Much of the rest of this water, though, would pass through a closed siphon, passing from the slopes of San Miguel, through the pueblo, and then to the uppermost points on the pampa. Thus the pampa, instead of having irrigated lands only below its 3,400 masl. rim, would be completely irrigated from outlets at 3,477 masl. This would benefit the cooperative and its members enormously, but it would also create even greater inequalities and conflict between the cooperative and the community, which would not gain any new irrigated areas. It would also impede further work and development on ditches which would benefit the community, a larger organization, with more members, than the cooperative. Some of Plan Meris' personnel were keenly interested in local technologies and the possibility of complementing their more costly projects with others undertaken by local organization with local labor and materials. With sufficient effort and understanding of the multiple constraints and demands on community members' time and energy, it might be possible for Plan Meris engineers to provide the direction and materials necessary to stimulate and then organize community labor to rebuild the high ditch to the Qolla Wayku valley. Others of that agency, however, tended to be somewhat condescending about what they considered to be primitive techniques on an unprofitably small scale. Which set of attitudes prevailed during the construction of ditches and subsequent agricultural development programs would importantly affect the extent to which potential inequalities and conflicts could be avoided and sufficient community organization to maintain both new and old systems could be developed.

If, though, Plan Meris did not complete its project, Huanoquiteños would either have to organize themselves in ways to help them maintain old ditches and build new ones, or face an eventual collapse of the environmental modifications they received from their forefathers and on which their livelihood now depended. We have seen thus far that a succession of exogenously imposed, sometimes brutally coercive regimes, abrogated cultural systems conducive to the combination of broad-gauged apprehension of landscape as system with detailed understanding of the material and hydrological characteristics of the multiple places within it. We have also seen how this sequence of regimes has moved the locus of labor-mobilizing authority further and further from its earlier bases in consensual acceptance of local, knowledge-based authority. Both of these losses make effective community initiative highly unlikely. More fully understanding how the greater environmental sensitivities and capacities of the earlier, larger community were achieved, and how they were lost, may give us clearer ideas about future possibilities, though. In the final chapter, therefore, I consider the sources of both individual environmental knowledge and of collective mobilization, and then analyze how both were destroyed. Both of those questions may help us to understand how Huanoquiteños' knowledge of their environment emerged, evolved, and withered.

Chapter Five

Cosmology, Society, and Nature

In this chapter, I recount some of the stories that helped me to interpret the visual and physical lessons I learned from looking at, moving along, and working on the ditches, as well as from the instruction and explanation I received about how to maintain and repair the ditches. Some of these stories are myths and legends recounting the interaction between earth and water in wild places; others are about *huacas*, material beings—often associated with specific sacred places, including places along *ceque* lines—that have specific supernatural or superhuman capacities and strengths; still others recount the interaction between individual humans and particular *huacas*, and many relate historical events involving named persons from Huanoquite and elsewhere. *Apus*, or mountain spirits, appear in many of them. Some of the most powerful stories recount gifts to and curses on Huanoquiteños from specific animals. Others relate how communication with and respect for earth, water, and sacred animals enabled their ancestors to alter landforms and water courses in ways contemporary Huanoquiteños cannot. I heard many of these stories as responses to my questions about ditch construction, maintenance, and collapse.

These stories emerged, and their meanings became clearer, in the context of the physical and visual work I was doing to trace and understand the paths of the ditches. In the following section, I recount how my explorations led me to, and helped me to understand, the most powerful and consequential of these stories, which revolve around the snake with golden braids, who controls the irrigation water in the ditches and in the rivers that feed them.

Finding the Snake with Golden Braids

My initial interpretation of these stories was that they manifested lost technology and lost awareness of the environment. My subsequent examination of the shorter ditches designed around particular, usually larger than average, pieces of private land showed this interpretation to be too simplistic. The private ditches were labor-intensive, imaginative devices based on a clear understanding of the proximate environment and of hydraulic technology. They were also the only ditches being built, and certainly the only ones that were carefully maintained.

They differed from the old ditches in that they were oriented to small parcels of private property. Therefore, they did not require, or attract, the huge collective effort that went into the old ditches. They did not demand the complexly interconnected topographic and technological knowledge and imagination that the old ditches had, so did not engender the combination of technological genius and precise understanding of complex relations of hydrology and topography that the old ditches did. Knowledge of the specific local forms that configured water and earth, as they occurred naturally and as they might be manipulated socially, across the entire landscape of Huanoquite's drainage basin, and of the contiguous drainage basins that contained Llaspay and Vilcabamba, was no longer relevant to the construction of ditches. A political system where the dominant property regimes were oriented to fragments of privately held land provided no reason to envision social engagements with and manipulation of telluric and hydraulic dynamics for the region as an integrated landscape.

Individualization of land tenure enhanced the ways that the superimposition of two exogenous regimes of land tenure and use immobilized community labor. As we saw earlier, two incompatible, nationally mandated, legal and bureaucratic regimes—one based on the 1930s community law, and the other on the cooperative legislation incorporated into the 1970s agrarian reform—regulated the two collectively owned properties in Huanoquite, i.e., the Community of Maska and the Tihuicte Cooperative. Each regime separately a) generated inequalities between putatively equal members, b) created tensions between individual and communal rights and obligations, and c) did not provide social mechanisms that ensured mobilization of members in collective tasks or due care of irrigation infrastructure. The territorial division of the properties exacerbated the social organizational inequalities within and incompatibilities between the two regimes. Most of the community's land lay on mountain slopes and upland valleys on the higher end of the ditch, while the cooperative owned the pampa, on the lower end of the ditch. The division of interests that emerged from these different spatial connections to the ditch were aggravated by very different organizational principles. The logic and ideology that the community maintained from its autonomous, indigenous roots was expansive and inclusive, while the logic and ideology of the cooperative, based on shared participation in a unitary private property, was exclusionist and bounded.

These multiple social and spatial sources of fragmentation prevented the kinds of individual vision that had seen opportunities to construct unified irrigation systems across the broader landscape. They also impeded collective response to the material and spiritual challenges of realizing those opportunities. The imagination and practice—topographic, technical, and social—required for these shorter, personal ditches no longer responded to the challenge of maintaining maximum elevation to reach as broad a community as possible. Just as there was no reason for the builders of minor ditches on private land above the pueblo to develop the sophisticated topographic and technological imagination of their ancestors, there was no reason for the particular owner of private land near the source of the Qewar to think about getting water across the ridge at Pukaqocha, even though he was using the ditch that the ancestors had built to do just that. Nor did he have any reason to change his own, or others', beliefs about the fate, or the intention, of the snake. The snake that embodied community and its intimate involvement with earth and water was no longer relevant to modifying the landscape, and so was relegated to a misremembered history.

Eventually, I realized that the snake was both the ditch and the water in it, and in this sense articulated the naturally determined course and behavior of the water in the ditch with the ditch's absolute social dependence on continued and persistent human maintenance. The snake, and the water, thus required continued closely coordinated solidarity, authority, and effort within the extended community that the ditch created between the widely dispersed hamlets and villages that used and maintained it. Killing the snake was thus the Spanish suppression of the local leaders whose special communication with and understanding of the snake, and of the spirits of earth and water, enabled them to create and preserve the ditch's technology and to inspire and organize the community *faenas* that kept the snake alive.

The snake's curse against anyone who might use the old high ditch affirmed that, without precise, locally grounded, technological/topographic knowledge, and without broadly mobilized community efforts to implement it, the ditches would pass from their benevolent provision of water into a deadly channeling of water to the weakest points in the ditch. These points were usually above the steepest slopes most susceptible to devastating erosion and loss of crop land, so that breaks in the ditch were potentially fatal. To the extent that it was precisely the technological and topographic knowledge, together with the broad mobilization of collective effort, that engendered, sustained, and ultimately constituted the community, the snake was not only wild water in the rivers and tame water in the ditches; she was also community. As community, she linked society and nature in their participation in and dependence upon each other.

Like wild water, the snake could flow both within the earth and along its surface. Also like wild water, the snake emerged from and was shaped by land, but her movement across the land transformed the landmasses (compare Zuidema, 1964:50; Salomon, 1991:45 fn). As tame water in the ditches, though, she mediated between society on the one hand and the dynamic, mutually transforming interaction of earth and water, turning their natural, extraordinarily munificent but also extraordinarily dangerous powers into resources that humans could safely use if they applied themselves diligently to caring for her and her ditches. She was thus also the topographic and hydraulic awareness of the entire landscape that simultaneously supported and was nurtured by the extended community, and the authority of the leaders who commanded and regulated the community's fulfillment of their obligations to her (compare Salomon, 1991: 18 on the authority of water-priests in Huarochiri and Gelles, 2000:88 on the staff of the two-headed snake as linking Cabaconde's water mayors both to the earth mother and to the mountain that provided that community's water).

In the most basic sense, therefore, it was not technology that was lost, but rather community, and community's relation to landscape. The loss of community abrogated both the opportunities and the challenges of building and maintaining a shared hydraulic system that embraced a variety of ecological zones within a unified landscape. What my informants related as the snake's curse was not simply their own individual death, but the death of the extended community. It was also the death of a particularly intimate and expansive engagement of community with nature through the landscape.

Animals, Earth, and Water in Incaic and Pan-Andean Cosmologies

The multiple interconnected social and natural attributes of the snake that linked the community of Huanoquite and its ditches to the river that flowed underground from the high plain of Qaranqa resonate with Salomon's (1991:14-16) discussion of how "the Huarochiri people called the world and time together *pacha*, an untranslatable word that simultaneously denotes a moment or interval in time and a locus or extension in space....The same word *pacha* is also the name of the earth in general." Salomon's discussion of the earth and its predominantly female associations shows that it is pervaded, bounded, and shaped by water and its predominantly male associations. "Water—rainstorms and mudslides, snow and glacial runoff, tiny irrigation canals and mighty rivers, even the astral river we call the Milky Way—is the kinetic part of the world." As storm, as downward flowing water, and as mudslides, this mobile male force cut into and transformed earth.

The "hydraulic embrace of moving water and enduring earth. . . yielded a biotic system in which life forms emerge from mixed earth and water. . ." As the depth, solidity, and stability of earth mixed with the fluidity and movement of water in "the circulation of water over soil...lives that are both watery and earthy emerge; plants, animals, people." In their notes to the first chapter of the Huarochiri manuscript, Salomon and Urioste (1991:44-45 fn) recount that Cuni Raya, a trickster deity who makes and remakes the world, is primarily a water deity, creator of irrigation, and able to create springs and rivers, and thus "associated with the hydraulic transformation of landforms."

Salomon (1991:4) warns that one must appreciate that "the equation between Inca religion and Andean religion is an ideological sleight," though he also recognizes Pan-Andean patterns of explanation and belief. He goes on to distinguish between "Inca invocations to the unseen source of order and beauty" on the one hand and local, once subordinate, populations' worship "focused on sacred beings peculiar to particular kin groups, villages, mountains, canals, and so forth." Salomon later elaborates the centrality of these localized sacred beings in Andean religion; "The horizon, not the cosmos—geography, not metaphysics—poses the question to which its most vibrant deities give answers. Andean numina lodge in places or placed objects; mountains, springs, lakes, rock outcrops, ancient ruins, caves."

These highly localized and materialized deities are called *huacas* in the diverse local Andean religions (Zuidema, 1964; Bauer, 1998), but Salomon may overdraw the distinction from Inca religion, which identified and used *huacas* in the same way. Local religions may not have identified an invisible Vira Cocha as a prime mover; but the highly abstract concepts about water's and earth's material and physical aspects as conceptually distinct forces in both Huanoquite and in Huarochiri, and the complex notions of how they interacted to form each other and multiple life forms, may be as profoundly cosmological and metaphysical as the more precisely elaborated (and more fully documented) religion of the Inca. Sherbondy (1982:24) states that "ancient Andean peoples not only possessed practical knowledge concerning subterranean hydrology, developing techniques for using these subterranean waters, but also elaborated a cosmology based on this knowledge."

These cosmological understandings all are rooted in local material processes and configurations, in Huanoquite and elsewhere. Allen (1988) notes Andean explanations of how water underground can flow against the direction of the surface water. Huanoquiteños understandings of how water flows from Qaranqa and under the high ridge of Vilcacunca to emerge as the Qewar's gushing source involve similarly elaborate, but locally rooted, conceptions of abstract, and thus empirically unobservable, general forces and of the material processes they regulate, some of which cannot be directly experienced because they occur deep in the earth. The ability to flow in different directions enables the circulation of water, and of the snake, that sustains itself and life as an ongoing cycle though multiple forms—from flow to vapor to cloud to raindrop to flow—in, on, and above the earth. Gose (1994) notes the cosmological significance of this bi-directionality of water in the Andes in his analysis of the intricate and consequential interactions between water, earth, seed, sowing, fertility, life, death and sex in the rituals and the beliefs related to irrigation in Huaquirca, a town south of Cuzco. Because irrigation systems must respond so precisely to local topography and hydrology, these cosmologies are clearly local in their origins and specifications. The site-specificity of their formulation, thus, in no way reduces their account of the general properties of water and earth, or of how these elements interact in the creation and shaping of life.

Salomon implies polar distinctions between horizon and cosmos, between geography and metaphysics, between concrete and abstract, and between place-specific and general. Complex cosmologies, though, can be rooted in specific, materially differentiated, sites and sacred places. The snake traced the course she would follow as water in the ditches through innumerable places, each differently formed by the interaction of earth and water. These places, though materially differentiated in space, were all connected within a system of vertical and horizontal relations, and within an order whose sequence was essential to the ditch's union of earth and water in order to create labor, fertility, and reproduction. These relations both depended on and contributed to the reciprocal formation of earth and watercourses in the landscape that molded, and was molded by, these elemental forces.

Some of the places where the snake led the ditch were made powerful and sacred by the ways that water, air, and earth met there; and the power of these specific places gave form to the abstract forces that drove the cosmology. Similarly site-specific, or place-derived, cosmologies are evident in stories about place that Basso's (1996) Apache friends told him while they taught him that "wisdom sits in places." They are evident as well in what the Kaluli of New Guinea told Steven Feld (1996) about how air moves through and communicates with earth, water, plants, animals, and air in ways, and with sounds, it reproduces as it passes through their own bodies, spirits, and voices as song and as speech.[1]

The Apaches, the Kaluli, and the Huanoquiteños participate in an ongoing interaction between abstractly constant attributes of society and nature. These attributes configure differently in specific places and times, and their configurations can be radically altered by cataclysmic social or natural events, ranging from conquest and rebellion to earthquake or storm. The histories that explain these events, though, are still formed by the abstract invisible constant forces, and by the material regularities they govern, even while they are formed by the actions and powers of individuals and communities with distinctive political, technological, artistic, and cosmological styles.

In the Andes, as in the Amazon, these communications with and about place, and between place and the natural but invisible forces of earth, air, water, and fire that make each place up, are importantly mediated by animals. The Huarochiri manuscript that Salomon analyzes describes ongoing dependence of earth-bound specimens of particular species of animals on specific constellations that actively sustain each species' characteristics and capacities. Salomon relates that Andean shamans gain knowledge and power by communicating with, and by assuming the species capacities of, particular animals. As Reichel-Dolmatoff (1976) tells us about Amazonian shamans as well, this knowledge and power constitute their ability to mediate between their society and the nature in which it participates. Their power and knowledge enable them to devise technologies that make more productive and at the same time conserve the landscape that society and nature combine to create. For the Huanoquiteños, as for the Tukano and Desana whose cosmologies Reichel-Dolmatoff describes, animals inform and instruct especially sensitive human individuals about how their community can engage air, earth, stars, and water— both directly and through the varied life forms they create—in ways that sustain and preserve human life together with the non-human life in which it participates and on which it depends.

All members of these various species receive and are sustained in their species essence by generative constellations that crystallize the essence or ideal of that particular species. The snake, interacting with landscape and community as they formed ditches in Huanoquite, shows that the *huacas* are the locally and

1. See Feld and Basso, eds. (1996) for other examples of place-based cosmologies in a variety of cultures.

materially specific configuration of the general forces and the spirits of water and of earth. These more abstract forces and spirits interact and form anew particular places constructed by previous interactions of water and earth. She, with other animals and spirits, instruct and regulate human activity in ways that enhance community well-being while protecting the systemic integrity of the landscape. The snake thus mediates between abstract general forces and their concrete locally specific manifestations. She also communicates the systemic relations between abstract general and locally specific processes. These complex and ongoing interactions constitute simultaneously a metaphysics and a geography. It is impossible to know which is prior or finally determinative, but both converge in the history of events, social and geological.

The snake was not the only animal that protected earth and water while communicating its secrets to particularly sensitive people. As I followed the old ditch and puzzled over how it had crossed the Qolla Wayku, I encountered another animal whose role helped me to understand more about the snake. I learned that a huge duck guarded the place where the ditch had crossed the river, assuring by the threat of death that no unauthorized person would tinker with the ditch or with its precious supply of water. I was working to reconstruct the vast knowledge of landscape, hydrology, material composition of subsoil, topography, and slope that would have been required to build the ditches high enough to include all of the lands of the various, spatially dispersed communities the ditches served. I could not, though, imagine how these ancestors had been able to learn the landscape so well that they could visualize an irrigation system that sustained the constant drop needed for flow with the minimal drop needed to maintain altitude, until I understood how snake and duck worked together. The snake and the duck—one absolutely bound to sinuous, flowing movements within and upon the earth's surface, the other able to soar over the landscape, see its broad contours from above, and then return to water—together had the physical and visual perspectives necessary for a full comprehension of all of the sequential places along the ditch, their relation to all the other places on the ditch, and the systemic relation of the ditch to its highly varied landscape and to the varied territories of the multiple hamlets and villages the ditch would serve. The snake and the duck thus incarnate how the sensations of moving through and between places, and of experiencing kinetically the articulation of places within a landscape, constitute essential ingredients of all eco-cosmologies, and of the dynamic interactions they enable between society and nature. These particular ingredients are especially critical in cosmologies that explain abstract principles and prescribe ethical behavior in mountain irrigation systems.

Thus, the duck and the snake embody the special knowledge that the designers and organizers of the ditch possessed. The two animals combine a) the general principles of hydrology, gravity, preservation and transformation of matter, inertia, weight, and the continuity of place in space that absolutely and unchangeably determined the forces of sustenance and decay that ruled the ditch with b) the regular annual processes of seasonal change, agricultural calendars, and alternating repair and use of the ditch, and with c) the individual human stories and the wildly changing social histories of the ditch's human users.

The ability to imagine the ditches' courses across the landscape, to understand how to build them in ways that would support the weight of rushing water moving almost horizontally across steep vertical slopes on surfaces that ranged from solid volcanic rock to highly friable and unstable mixtures of earth and broken rock, and then organize and train the vast labor force required to implement the solutions to these material and spatial problems in ways that conformed to their individual and community rights, interests, desires, and variable social powers required an integrated awareness of and action within three different, but deeply articulated, levels of reality and meaning—a) abstract, unseen, and unchanging physical laws, b) the observable material process that these abstract and unseen laws govern, and c) the historically contingent, partially unpredictable, highly variable, and constantly changing human agencies. The community gave authority and respect to those who combined and worked within these three kinds of awareness. The conquering Spaniards extirpated them in fear that their power to mobilize labor might also enable them to mobilize resistance. This suppression catalyzed a self-reinforcing chain of loss—of community identity, of ecological integrity, of technological and environmental understanding, and of eco-cosmological meaning.

The snake linked the most difficult, and thus most charged with social history, places on the ditch. These were places that had required the greatest creative leaps of understanding and the most arduous social efforts in order to build and maintain the ditch over or around them. The most difficult of these places, where the old ditch had been cut into a vertical rock wall, was located on a *ceque* line (cf. Zuidema, 1964) that linked Ccori Qancha, a sacred site in Cuzco, to a site where a powerful underground flow of water from a sacred lake high above the Apurimac river could be heard. In other words, the snake was participating simultaneously in natural, social, and sacred domains. Finally, it was clear that the snake in some sense depended on the community for its existence and survival, though it took me a long time to understand how or why a wild animal would depend on humans, or how this same wild animal mediated between humans and the natural world. The snake not only embodies community efforts in building and maintaining the ditch to protect the waters in it: the snake also mediates between the wild—or natural—state of the water flowing underground and the social condition of the water in the ditches.

Two points on the ditch were particularly critical to the snake's need for community to build the ditches in which she would flow. The ditch was able to reach both of these places only because it had had been built with minimum drop across very steep slopes and sheer rock faces. By traversing these ridges, though, the ditch was able to continue into entirely other drainage systems. This allowed it to flow over high topographic barriers, and then on to an additional network of communities below. The ancients' ability to do this created, but also depended on, two distinct cognitive and cosmological achievements: first, knowledge of the landscape as a system of linked places whose relations to each other enabled the community to construct ditches that would pass the water from one drainage basin to another; and second, the formation of knowledge-based, technically competent, labor-coordinating principles of authority and community.

Lessons about Nature's Society and Society's Nature

The myths, legends, and stories that explain the ways that landscape, technology, and community mold each other in Huanoquite suggest useful ways to think about the ways that different cosmologies explain the relations of society and nature. Histories of the snake, the duck, and the ditches integrate all three of the levels of reality: a) abstract general forces, b) specific material processes, and c) narrative history that I discussed above. They specifically link the narrative of contingent events with catastrophic consequences, i.e., the Spanish conquest and the imposition of the hacienda regime, to the abstract principles and material processes of the community's relation to landscape. They explicitly associate the snake's curse and the duck's menace with the Spanish conquest and with the destruction of the social organization needed to maintain the oldest and highest ditch. The Spaniards' zealous campaign against indigenous local leaders, whose systematic torture and assassination they justified by calling it the "extirpation of idolatries," would have included among its targets those indigenous authorities and leaders responsible for organizing ditch maintenance. Such powers would be manifest both in their special knowledge and in the authority through which they mobilized community effort. The special knowledge that enabled these indigenous authorities to fit collective action to the material processes of their environment on the one hand, and to the beliefs and practices of their fellows on the other, died with them. This special knowledge was as essential to the survival of community as it was to the community's ability to mobilize the care, precision, and labor required to sustain the ditches' positive contributions to agriculture and to prevent the devastating damage they could cause if they were not kept intact.

Local stories still recount the "visitas," the investigations by colonial clergy into local conditions and cultures through which these extirpations were organized, as forces that ended the powers that their ancestors had over the movements of non-human animals and objects. When the Spaniards suppressed the local authorities that organized ditch maintenance, the ditch collapsed and the snake retreated. The broader community was left to retreat as well, leaving isolated hamlets dependent on, and therefore even more subordinate to, the exogenous *hacendados*. The snake, embodying the physical and topographic reality of the ditches, encompasses not just local community, but the extended community that linked multiple places across several drainage systems before the coming of the Spanish and the dominion of the *hacendados*.

Much later, I followed the snake in ways that helped me see that she had made possible the union of multiple hamlets into a single community, and that this extensive community collapsed into separate, often antagonistic, social entities when the ditches collapsed. Thus, while the snake embodies trans-temporal relations between earth, water, and human community, she also reacts to precisely the kind of exogenous cataclysm—i.e., conquest by the Spanish and subsequent neglect by the Huanoquiteños—that Viveiros de Castro (1996) claims the attribution of naturally autonomous reality to some element represented in a cosmology would make impossible. Though rooted in hydro-thermal geology, the timeless snake in fact watches carefully two very different levels of temporal reality: a) the annual calendar and the astronomical changes that not only mark seasons that men must take into account to schedule their agricultural tasks, but that also enumerate the time required for each of the multiple specialized material processes—floral and faunal—of insemination, gestation, and birth and b) the cataclysmic social and geological events that constitute natural and social history.

I met the snake while working on the more recently built ditch to the pampa that is still in use today, but all of the stories about the snake involve the much higher, now abandoned ditch that once embraced so much of the cultivable parts of Huanoquite's drainage system. The newer, still used ditch, built within the last 200 years, does not have anything like the mythic significance or richness of the older, abandoned one. I heard no stories about animals guarding parts of its trajectory. It clearly contributed enormously to the agricultural prosperity of the villagers, and there was general concern about its maintenance, as well as discussions and complaints about the lack of coordination in its maintenance, exacerbated by competition over the water, between the community and the Tihuicte cooperative. My questions about ditch maintenance catalyzed these stories and criticisms, as well as a general observation that since the agrarian reform of the 1970s, everyone had to attend to the full range of commercial agricultural activities, including trips to Cuzco to sell crops and to buy implements. The point of many of the stories I heard about contemporary problems of ditch maintenance was that the *hacendados* had done these things before, and that one of the consequences of the reform was "no nos alcanza el tiempo" to perform the tasks of maintaining ditches and terraces that were essential to the long term ecological integrity and productivity of the area. Again, though, as I listened longer and more carefully, there was ambivalence in these stories and complaints linked to the memory that the construction, maintenance, and use of the lower ditch had been at the behest, and under the supervision, of a small group of oppressive land owners. There was no such ambivalence in the stories about the authority that had coordinated building and maintenance of the older ditches. Rather, the Huanoquiteños saw their ancestors from that time as magically powerful and as socially and technologically beneficent. In consequence, nature was much more munificent.

Irrigation, Knowledge, Authority, and Power

These historical memories cast severe doubt on the notion that irrigation systems necessarily engender authoritarian or despotic rule, or even that they require a tightly controlled or coerced labor force. Wittfogel, and the Western social scientists who make similar arguments, assume that the intense quantities of labor required to make significant infrastructural changes in landscape can only occur under an autocratic, authoritarian, or absolutist regime. They may, though, be mistaking path-dependent patterns in European political history, deriving back to Egyptian and Grecian antecedents, for universal rules. We may need to consider how firm our evidence is that the road, storage, and irrigation systems in the Andes indicate powerful central authorities. Indeed, some archaeologies and histories conclude that the brief Incaic imperium essentially coopted existing small and relatively egalitarian but intensely cooperative communities that had already modified their own landscapes in ways that enabled them to significantly intensify agricultural production. These same studies indicate that the Incaic empire depended on extensively decentralized rule that continued with existing local *ayllu* structure and administration. In this, they are essentially replicating Clements Markham's (1872, 1964) 19th century admiration for Incaic administrative ingenuity and economy in finding ways to allow local systems to continue to operate within a loosely connected but materially effective system of rule.

Close attention to the materio-spatial characteristics and social organizational requirements of mountain irrigation systems reinforce this alternative interpretation. The complex interdependence of natural and social dynamics within the precise constraints of physical laws and rules configured differently in multiple, systemically articulated sites make it almost impossible for standardized central authority to dictate how an effective ditch system should be built and operate. Indeed, what we have considered so far of Huanoquite's ditches convinces me that extensive physical, cognitive, detailed and systemic knowledge of the natural landscape, its material composition, multiple specific places, the topographic and hydrological relation between its different places, and the characteristics and locations of the best agricultural land in relation to these other places would be absolutely essential to the imagination needed to design and build the ditches, and that this kind of knowledge would only be available to a particularly sensitive individual with long-term experience of this particular landscape. Exogenously imposed authority with anything like absolute power seems incompatible with such locally effective authority. Indeed, the collapse of agricultural intensity under the Spanish absolute rule affirms that the Incaic system was far more open to locally specific authority and knowledge. From this, I infer that the authority of local ditch builders over construction and maintenance was knowledge-based, not power-imposed, and that their ability to mobilize and coordinate the labor needed for the complex, urgently punctual maintenance of ditches built across slopes—from steep to sheer—on difficult surfaces that ranged from highly unstable soils to solid rock faces was earned through their understanding rather than imposed by force of arms.[2]

For all Huanoquiteños, there is magic and danger, as well as blessing, in water. The Qewar, with its powerful spring out of the ground and its dramatic final fall, is especially potent. Deadly spirits protect both the Qewar's fall at Paqpiri, where people have been said to die if they dared stay near it, and the major waterfall in the Qolla Wayku's gorge, just below where the ancient high ditch once crossed it. A huge duck also guards the place near the carved thrones

2. Trawick (2003) reaches similar conclusions.

where the Inca road crosses the Watanay. Sacred animals, and spirits of earth, air, and water, are all empowered and regulated by their relations to constellations and to the calendrical patterns they maintain. Their powers provide energy and strength to humans and to the non-human beings on whom society depends. They impose general, constantly valid, rules and principles on the being and on the behavior, not only of humans, but also of animals, plants, rain, rocks, and rivers.

Huanoquiteños believe that their ancestors derived from these animals, and from the earth, special strength and powers that they no longer have; that the ancestors were able to command stones and water to do their bidding. Some believe that their ancestors built tunnels to direct the underground water flows on which their agriculture still depends. Stories about mountains, rivers, rocks, and animals, and about ancestors and gods with special powers, explain and enforce abstract principles of how human and non-human beings behave. These stories also expand to include narrative accounts of social events and processes that had cataclysmic consequences. These stories all recount how the exogenously imposed historical changes reduced not only the power and productivity of the landscape, but also Huanoquite's ability to support that power and productivity. Conquest, subordination, and religious persecution disrupted social interaction with nature, precipitating ecological disruption and a loss of social meaning and capacity. This undermined the Huanoquiteños' sense of themselves and of community, leaving only the memory of their ancestors' powers.

The ditches and the stories remembered about them and about the places they traverse, do not simply affirm Parker's suggestion that significant landscape-changing technological innovations emerge from particular individuals' heightened sensitivity to their environment. They also tell us about the prelinguistic, meta-cultural kinds of communication that such individuals may sustain with earth, water, air, and fire. Finally, they announce to us that sustainable use and maintenance of these technologies may also depend upon the continued survival and authority of these individuals. We humans face the paradox that effective technology depends upon special sensitivity to the environment, but the careless, negligent, or greedy use of that same technology may seriously damage the landscape it was invented to make more productive.

More broadly than showing how effective technological innovation emerges from communication with earth, water, fire and air in ways that are simultaneously sensitive to abstract laws and regularities, to the material processes that manifest them in space and time, and to the historically contingent formation of culture and social organization, the ways that the snake mediates between nature and society in a specific site—Huanoquite—opens vistas for understanding how Huanoquite fits into far larger systems, and how Huanoquiteños' cosmologies take account of broader participation. The east-facing mountains that mold the dynamic and constantly moving embrace between earth and water that the snake embodies were uplifted many thousands of years ago in ways that reversed the westward flow of waters, and created instead a barrier to the onward flow of clouds that cycle up the Amazon, and that then move up the Apurimac to Huanoquite. These mountains now stop the clouds, chill them, and turn them to rain. It is this rain that ultimately feeds water to Qaranqa, as well as to the rest of the region's lakes, streams, and rivers, that ultimately return it to the Ccorca, which feeds it to the Apurimac, then the the Amazon, and finally to the sea, where the water circulates back up into clouds and moves back toward the Andes again. Ecologically attentive Huanoquiteños can see these clouds coming from the eastern lowlands into which the Apurimac flows, and know that the water returns there, often with heavy loads of soil, after it falls as rain. Snakes embody water's cycle from rain, to flow right above and within earth, to rivers molding land and societies in the Andes and in the Amazon.

The snake that mediates between water and land is of a size and strength that only the tropical rivers below could support, and in that sense must come from there. Lathrap (1968, 1974, 1977) has argued that the incorporation of iconography representing Amazon water animals—particularly snakes and caymans—into Andean obelisks and other monumental art indicates that the Andes were settled from the Amazon as various populations exceeded the carrying capacities of the different mouthbays where they had developed technologies and social organization that combined agriculture, hunting, and fishing. The direction of population movement, though, is not necessarily related to the sharing of cosmologies. Large snakes connect multiple villages and ecosystems in the Amazon, clearly embody the complex, changing, and amorphous relations between land and water there, reflect the fecundity of those relations, and guide humans in their engagement with those relations (Reichel-Dolmatoff, 1971, 1976; Whitehead, 1998; see also Hornborg, 2001).

These continuities from the Amazon to the Andes in the relations of society to nature strongly suggest the futility of attempts to describe these relations in linear, dichotomous, or unidirectional terms. Arguing about whether the environment shapes society or whether society constructs nature misses the essential lesson that society and nature interact from very different material bases, organizational forms, reproductive dynamics, conceptual frames and temporal systems, but that they do so in reciprocally formative ways.

These ways are necessarily specific to particular places. The kinds of interaction between society and nature possible and desirable in a water-scarce, high level, cold mountain environment respond to different challenges and to different opportunities than those of the lowland tropics. The ongoing discovery and amazement of Balée, Descola, de Castro, and others that extensive landscapes in the Amazon show traces of human modifications of the environment needs to take full account of how much communication between humans and nature, and how much natural systems manifest in particular places, influence both the imagination of technology and the regulation of its implementation.

Bibliography

Abram, David. 1996. *The Spell of the Sensuous.* New York: Vintage Books.

Alden, Dauril. 1976. "The Significance of Cacao Production in the Amazon Region during the Late Colonial Period: An Essay in Comparative Economic History." *Proceedings of the American Philosophical Society* 120 (Apr.): 103-135.

Allen, Catherine. 1988. *The Hold Life Has: Coca and Cultural Identity in an Andean Community.* Washington: Smithsonian.

Arguedas, Jose Maria. 1964. "Puquio, una cultura en processo de cambio." Pp. 221-272 in *Estudios sobre la Cultura Actual del Peru.* Lima: Universidad Nacional de San Marcos.

Arhem, Kaj. 1996. "The Cosmic Food Web: Human-Nature Relatedness in the Northwest Amazon." Pp. 185-205 in Descola and Pálsson.

Balée, William. 1994. *Footprints in the Forest: Ka'apoor Ethnobotany—The Historical Ecology of Plant Utilization by an Amazonian People.* New York: Columbia University Press.

Basso, Keith H. 1996. *Wisdom Sits in Places.* Albuquerque: UNM Press.

Bateson, Gregory. 1972. *Steps to an Ecology of Mind.* Frogmore: Paladin.

Bateson, Gregory. 1979. *Mind and Nature: A Necessary Unity.* New York: Dutton.

Bauer, Brian S. 1998. *The Sacred Landscape of the Inca.* Austin: University of Texas Press.

Bradby, Barbara. 1975. "The Destruction of Natural Economy." *Economy and Society* 4: 127-161.

Brush, Stephen B. 1977. *Mountain, Valley, and Field: The Economy and Human Ecology of an Andean Valley.* Philadelphia: University of Pennsylvania Press.

Bunker, Stephen G. 1985. "Topographic Perception in the Construction and Maintenance of an Andean Irrigation System," presented at the meetings of the Latin American Studies Association, Albuquerque.

Bunker, Stephen. 1987. "Ritual, Respect, and Refusal: Drinking Behavior in an

Andean Village." *Human Organization* 46, 4 (Winter): 334-342.
Bunker, Stephen, and Linda Seligmann. 1986. "Organización Social y Visión Ecologica en un Sistema de Riego Andino." *Allpanchis* 18, 27: 149-178.
Descola, Philippe. 1994. *In the Society of Nature.* Cambridge: Cambridge University Press.
Descola, Philippe. 1996. "Constructing Natures: Symbolic Ecology and Social Practice." Pp. 82-102 in Descola and Pálsson.
Descola, Philippe, and Gísli Pálsson (eds.). 1996. *Nature and Society: Anthropological Perspectives.* London: Routledge.
Downing, T., and M. Gibson (eds.). 1974. *Irrigation's Impact on Society.* Anthropological Papers of the University of Arizona 25.
Ellen, Roy. 1996. "The Cognitive Geometry of Nature: A Contextual Approach." Pp. 102-123 in Descola and Pálsson.
Escobar Moscoso, Mario. 1977. "Estudio Comparativo de los Valles del Urubamba y el Apurimac." Ponencia al III Congreso Peruano de Hombre y la Cultura Andina, Cusco.
Feld, Steven, and Keith H. Basso (eds.). 1996. *Senses of Place.* Santa Fe: School of American Research.
Fonseca M., Cesar. 1984. "El Control Comunal del Agua en la Cuenca del Rio Canete." *Allpanchis* 19, 22:61-73.
Gade, D. W., and R. Rios. 1972. "Chaquitaclla: The native footplow and its persistence in Central Andean agriculture." *Tools and Tillage* 2, 1: 3-15.
Geertz, Clifford. 1959. "Form and Variation in Balinese Village Structure." *American Anthropologist* 61: 991-1012.
Gelles, Paul. 2000. *Water and Power in Highland Peru: The Cultural Politics of Irrigation and Development.* Brunswick: Rutgers University Press.
Gose, Peter. 1994. *Deathly Waters and Hungry Mountains: Agrarian Ritual and Class Formation in an Andean Town.* Toronto: University of Toronto Press.
Harris, Marvin. 1977. *Cannibals and Kings: The Origins of our Cultures.* New York: Vintage.
Harvey, David. 1999. The *Limits to Capital.* London: Verso.
Hemming, John. 1978. *Red Gold: The Conquest of the Brazilian Indians.* Cambridge: Harvard University Press.
Hornborg, Alf. 1996. "Ecology as Semiotics: Outlines of a Contextualist Paradigm for Human Ecology." Pp. 45-62 in Descola and Pásslon.
Hornborg, Alf. 2001. "Vital Signs: An Ecosemiotic Perspective on the Human Ecology of Amazonia." *Sign Systems Studies* 20, 1: 121-151.
Hugill, Peter. 1994. *World Trade since 1491.* Baltimore: The Johns Hopkins University Press.
Ingold, Tim. 1996. "The Optimal Forager and Economic Man." Pp. 24-44 in Descola and Pásslon.
Isbell, Billie Jean. 1974. "Kuyaq: los que nos Aman." Pp. 110-152 in Giorgio Alberti and Enrique Mayer (eds.), *Reciprocidad y Intercambio en los Andes Peruanos,* Lima: IEP.
Kosok, Paul. 1965. *Life, Land, and Water in Ancient Peru.* Brooklyn: Long Island University Press.

Lathrap, Donald W. 1968. "Aboriginal Occupations and Changes in the River Channel on the Central Ucayali, Peru." *American Antiquity* 33: 62-79.

Lathrap, Donald W. 1974. "The Moist Tropics, the Arid Lands, and the Appearance of Great Art Styles in the New World." *Special Publication of the Museum of Texas Technological University* 7: 115-158.

Lathrap, Donald W. 1977. "Our Father the Cayman, Our Mother the Gourd: Spinden Revisited, or a Unitary Model for the Emergence of the Agriculture in the New World." Pp. 713-731 in C.A. Reed (ed.), *Origins of Agriculture*. The Hague: Mouton Publishers.

Lehmann, David (ed.). 1982. *Ecology and Exchange*. Oxford: Oxford University Press.

Mann, Michael. 1982. *The Sources of Social Power,* Vol 1. Cambridge: Cambridge University Press.

Marglin, S. 1990. "Towards the Decolonization of the Mind." Pp. 1-28 in F. Appfel Marglin and S. Marglin (eds.), Dominating Knowledge: Development, Culture, and Resistance. Oxford: Clarendon Press.

Markham, Clements. 1872. *Report on the Discovery of Peru*. London: printed for the Hakluyt Society.

Markham, Clements. 1964. *Narrative of the Rites and Laws of the Incas*. New York: Burt Franklin.

Mayer, Enrique, and Cesar Fonseca M. 1979. "Temas Agrários en la Cuenca del Rio Cañete." Lima: Impreso ONERN.

Meinig, D.W. 1993. *The Shaping of America: A Geographical Perspective on 500 Years of History,* Vol. 2. New Haven: Yale University Press.

Mitchell, William. 1973. "The Hydraulic Hypothesis: A Reappraisal." *Current Anthropology* 78: 25-44.

Mitchell, William. 1976. "Irrigation and Community in the Central Peruvian Highlands." *American Anthropologist* 78: 25-44.

Montoya, Rodrigo, M. J. Silveira, and F. J. Lindoso. 1979. "Produción Parcelaria y Universo Ideologica: el Caso de Puaio." Lima: Mosca Azul.

Moseley, Michael E. 1974. "Organizational Pre-adaptation to Irrigation: The Evolution of Early Water-Management Systems in Coastal Peru" in Downing and Gibson.

Murra, John. 1975. *Formaciones Económicas y Politicas del Mundo Andino*. Lima: IEP.

Netherly, Patricia. 1984. "The Management of Late Andean Irrigation Systems in the North Coast of Peru." *American Antiquity* 49, 2: 227-254.

Ossio, Juan M. 1976. "El simbolismo del agua y la representación del tiempo y el espacio en la fiesta de la acequia de la comunidad de Andamarca." *Actes du Congrés XLII International des Americanistes,* v.4: 377-396. Paris.

Palmer, Robin, and Neil Parsons (eds.). 1977. *The Roots of Rural Poverty in Central and Southern Africa*. Berkeley: University of California Press.

Parker, William N. 1991. *Europe and America in the Wider World,* Vol 2. Cambridge: Cambridge University Press.

Peirce, Charles S. 1877. "The Fixation of Belief." *Popular Science Monthly* 12: 1-15.

Rappaport, Roy. 1968. *Pigs for the Ancestors. Ritual in the Ecology of a New-Guinea People.* New Haven: Yale University Press.

Rappaport, Roy. 1979. *Ecology, Meaning, and Religion.* Berkeley: North Atlantic Books.

Rappaport, Roy. 1994. "Humanity's Evolution and Anthropology's Future." Pp. 153-166 in R. Borofsky (ed.), *Assessing Cultural Anthropology.* New York: McGraw-Hill.

Reichel-Dolmatoff, Gerardo. 1971. *Amazonian Cosmos: The Sexual and Religious Symbolism of the Tukano Amazonian Indians.* Chicago: University of Chicago Press.

Reichel-Dolmatoff, Gerardo. 1976. "Cosmology as Ecological Analysis: A View from the Rainforest." *Man* 2: 307-318.

Rival, Laura. 1996. "Blowpipes and Spears: The Social Significance of Huaorani Technological Choices." Pp. 145-164 in Descola and Pálsson.

Rivera, José A. 1998. *Acequia Culture: Water, Land, and Community in the Southwest.* Albuquerque: UNM Press.

Ross, Eric. 1978. "The Evolution of the Amazonian Peasantry." *Journal of Latin American Studies* 10: 198-218.

Salomon, Frank. 1991. "Introductory Essay: The Huarochiri Manuscript." Pp. 1-40 in Salomon and Urioste.

Salomon, Frank, and George L. Urioste. 1991. *The Huarochiri Manuscript: A Testament of Ancient and Colonial Andean Religion.* Austin: University of Texas Press.

Scarborough, Vernon. 2003. *Flows of Power: Ancient Water Systems and Landscape.* Santa Fe: School of American Research.

Seligmann, Linda J. 1987. "Land, Labor, and Power: Local Initiative and Land Reform in Huanoquite, Peru." Ph.D. dissertation, University of Illinois, Urbana.

Seligmann, Linda J. 1995. *Between Reform and Revolution: Political Struggles in the Peruvian Andes, 1969-1991.* Stanford: Stanford University Press.

Seligmann, Linda, and Stephen Bunker. 1994. "An Andean Irrigation System: Ecological Visions and Social Organization." Pp. 203-232 in William P. Mitchell and David Guillet (ed.), *Irrigation at High Altitudes: The Social Organization of Water Control Systems in the Andes,* Volume 12. Society for Latin American Anthropology Publication Series.

Sherbondy, Jeannette. 1982. "El Regadio, los Lagos, y Mitos de Origen." *Allpanchis* 17, 20.

Sweet, David. 1974. *A Rich Realm of Nature Destroyed: The Middle Amazon Valley* (1640-1750). PhD. Dissertation, University of Wisconsin, Madison.

Tilly, Charles. 1995. "To Explain Political Processes." *American Journal of Sociology* 100, 6: 1594-1610.

Trawick, Paul B. 2003. *The Struggle for Water in Peru: Tragedy and Comedy in the Andean Highlands.* Stanford: Stanford University Press.

Urton, Gary, and Anthony F. Aveni. 1983. "Archaeoastronomical Fieldwork on the Coast of Peru." Pp. 221-234 in A. F. Aveni and G. Brotherton, BAR International Series 174. Oxford: British Archaeological Reports.

Villanueva, Horacio. 1982. *Cusco 1689: Economía y Sociedad en el Sur Andino.* Cuzco: Centro de Estudios Rurales Andinos Bartolome de las Casas.

Viveiros de Castro, Eduardo. 1996. "Images of Nature and Society in Amazonian Ethnology." *Annual Review of Anthropology* 25: 179-200.

Whitehead, Neil L. 1998. "Indigenous Cartography in Lowland South America and the Caribbean." Pp. 301-326 in D. Woodward and G.M. Lewis (eds.), *The History of Cartography: Cartography in the Traditional African, American, Arctic, Australian, and Pacific Societies,* Vol. 2. Chicago: University of Chicago Press.

Whitehead, Neil L. 2003. "Three Patamuna Trees: Landscape and History in the Guyana Highlands." Pp. 59-80 in N.L Whitehead (ed.), *Histories and Historicities in Amazonia.* Lincoln: University of Nebraska Press.

Wittfogel, Karl. {1928} 1985. "Geopolitics, Geographical Materialism, and Marxism." *Antipode* 17: 21-72.

Wittfogel, Karl. 1957. *Oriental Despotism.* New Haven: Yale University Press.

Zuidema, Tom. 1964. *The Ceque System of Cuzco: The Social Organization of the Capital of the Inca.* Leider: E.J. Brill.

Zuidema, Tom. 1986. "Inca Dynasty and Irrigation: Another Look at Andean Concepts of History." Pp. 177-200 in J.U. Murra, N. Wachtel, and J. Revel (eds.), *Anthropological History of Andean Polities.* Cambridge: University of Cambridge Press.

Index

www.ingramcontent.com/pod-product-compliance
Ingram Content Group UK Ltd.
Pitfield, Milton Keynes, MK11 3LW, UK
UKHW012042150926
475560UK00001B/145

9 780739 111970